D0518140

HUNTING the HIGGS

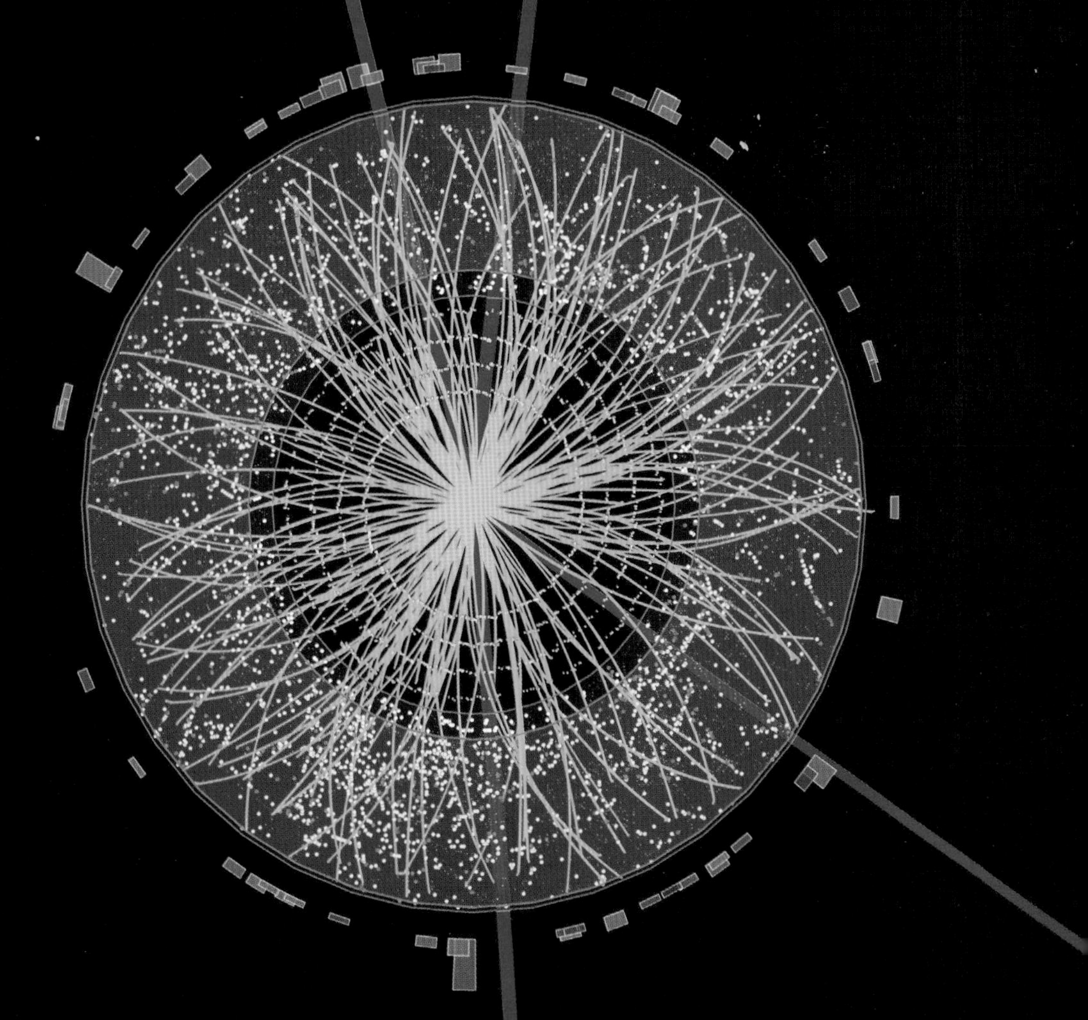

HUNTING the HIGGS

The inside story of the ATLAS Experiment at the Large Hadron Collider

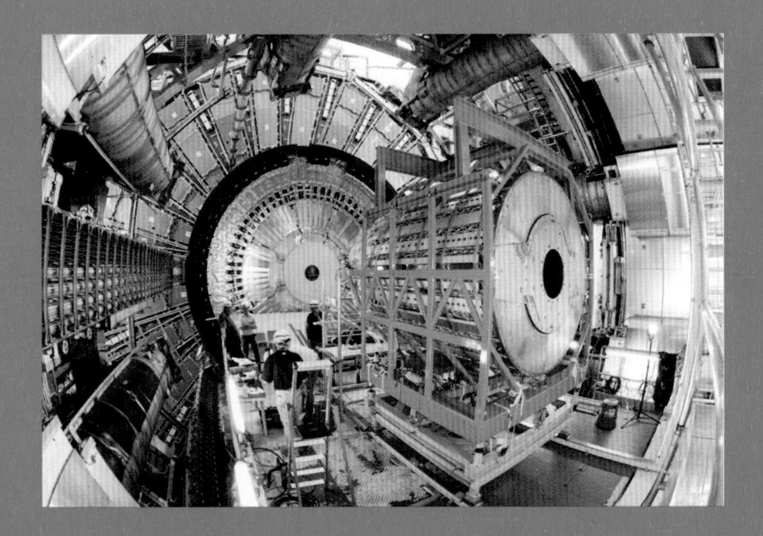

Claudia Marcelloni   Colin Barras

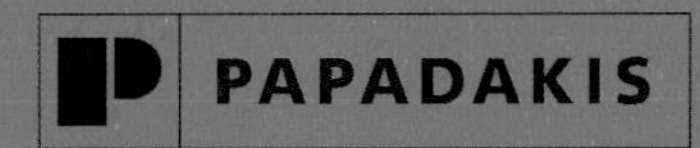

First published in Great Britain in 2013 by Papadakis Publisher

PAPADAKIS

An imprint of New Architecture Group Limited

Kimber Studio, Winterbourne, Berkshire, RG20 8AN, UK
info@papadakis.net | www.papadakis.net

@papadakisbooks PapadakisPublisher

Published in collaboration with ATLAS
www.atlas.ch

Publishing Director: Alexandra Papadakis
Design: Alexandra Papadakis
Claudia Marcelloni: photo editor / photographer
Colin Barras: text
Scientific Editorial: Peter Jenni, Fabiola Gianotti, Dave Charlton, Andy Parker, Daniel Froidevaux, Richard Hawkings, Beate Heinemann, Andreas Hoecker, Karl Jakobs, David Lissauer, Bill Murray, Aleandro Nisati, Thorsten Wengler

We would like to thank the ATLAS management for commissioning the work, and to all of the members of the ATLAS collaboration for their support in preparing it.

ISBN 978 1 906506 37 7

PHOTO CREDITS
All photographs by Claudia Marcelloni © ATLAS Experiment / CERN with the exception of: back cover: © ATLAS Experiment / CERN, Joao Pequenao, pp24, 32, 33, 34, 42, 43, 47, 50: Maximillien Brice © CERN, pages 10 and 12: © CERN, p26 © NIKEF, p27 – top © Argonne Nat. Lab and bottom left © Michael Schernau, p36: © Roy Langstaff, p58: © NASA, ESA, M. Jee and H. Ford (Johns Hopkins University).

A CIP catalogue record of this book is available from the British Library.

Printed and bound in China

Captions:
Front cover: Moving the end cap calorimeter during the January 2011 maintenance shutdown period.

Back cover: A proton collision event with 4 muons coloured blue.

Half title: A candidate Higgs boson decays into four muons coloured red.

Frontis: The giant eight-toothed end cap magnet after installation inside the ATLAS Cavern.

Title page: One of the end caps of the inner detector moves into place.

Contents

Introduction

If we allow ourselves, just for a minute, to pause and look back over the history of the ATLAS experiment, we might be surprised to learn that it already spans a quarter of a century! Initial plans to build one of the world's biggest particle detectors were made in the late 1980s. Manufacture and assembly took most of the late 1990s and early 2000s. Installation in the underground cavern started in 2003 and took almost five years. A scientific collaboration of unprecedented size assembled to build and use ATLAS - currently about 3000 physicists from 177 institutes in 38 countries, with more than 1000 students. Their dedicated and talented work, together with the crucial support from their respective institutes and governments, is what made one of the most sophisticated scientific instruments ever constructed a reality.

Thanks to the extraordinary performance of ATLAS and the Large Hadron Collider, the Higgs boson, for so long the most elusive of all subatomic particles, has finally made an appearance in a physics experiment, opening a new chapter in our understanding of the Universe.

More than that, the ATLAS experiment has shown once again how science can transcend geographical and political borders. In addition to being a truly international endeavour in itself, many tens of thousands of people from all over the world have participated in visits to ATLAS, to find out more about the work we do - either by coming to visit in person, or by a "virtual visit" from their school or university.

The voyage of discovery at the LHC has just begun. We expect that nature has plenty more surprises in store for us as we explore deeper into the subatomic world with the ATLAS detector. We hope this book will help you share in the endeavour and the excitement.

Fabiola Gianotti, Spokesperson, ATLAS Collaboration, 2009-2013
Dave Charlton, Spokesperson, ATLAS Collaboration, 2013-

One of the end cap calorimeters is moved into position using a set of rails. This calorimeter will measure the energy of particles that travel close to the beam when two protons collide.

Five, four, three, two, one...

At 9:30 am local time on 10 September 2008, the world's eyes were on Geneva. In the control centre at CERN, the European Organization for Nuclear Research, LHC project leader Lyn Evans flipped a switch and introduced a stream of protons into a stretch of the Large Hadron Collider. After nearly a quarter of a century of planning, the most ambitious physics experiment ever built had begun.

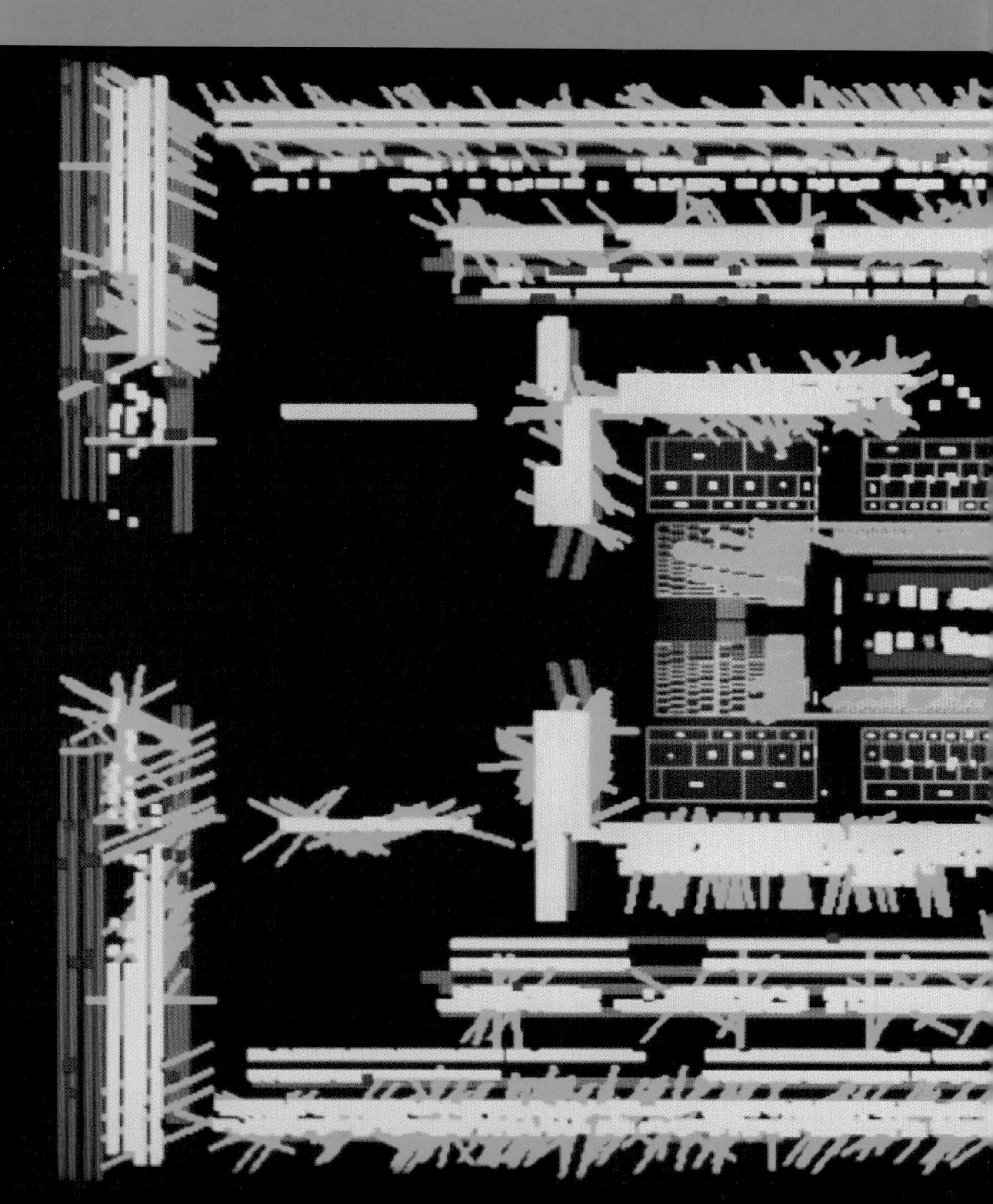

Zero!

The ATLAS control centre on LHC start-up day.

The ATLAS detector hums with activity as it sees a proton beam for the first time.

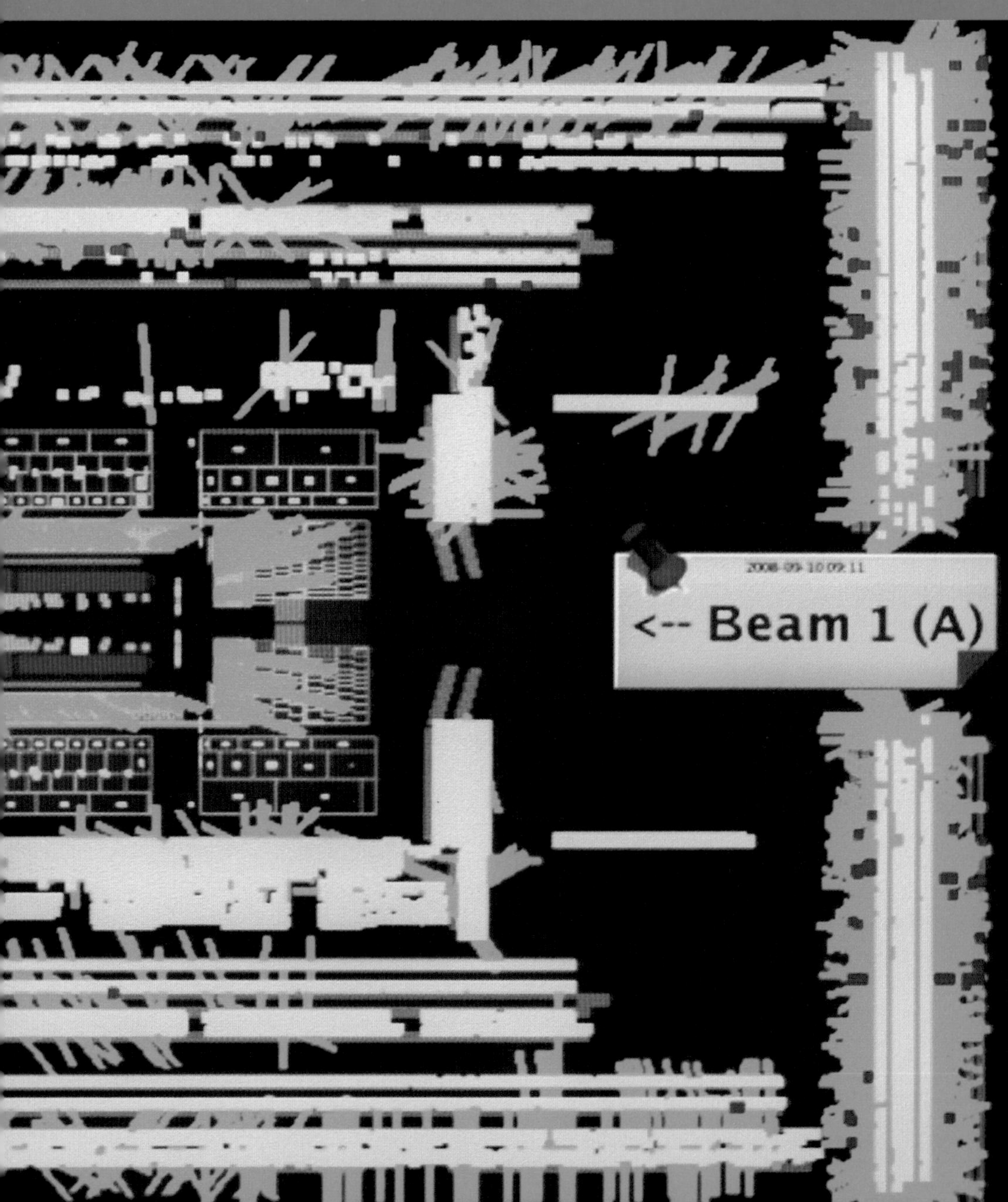

CMS

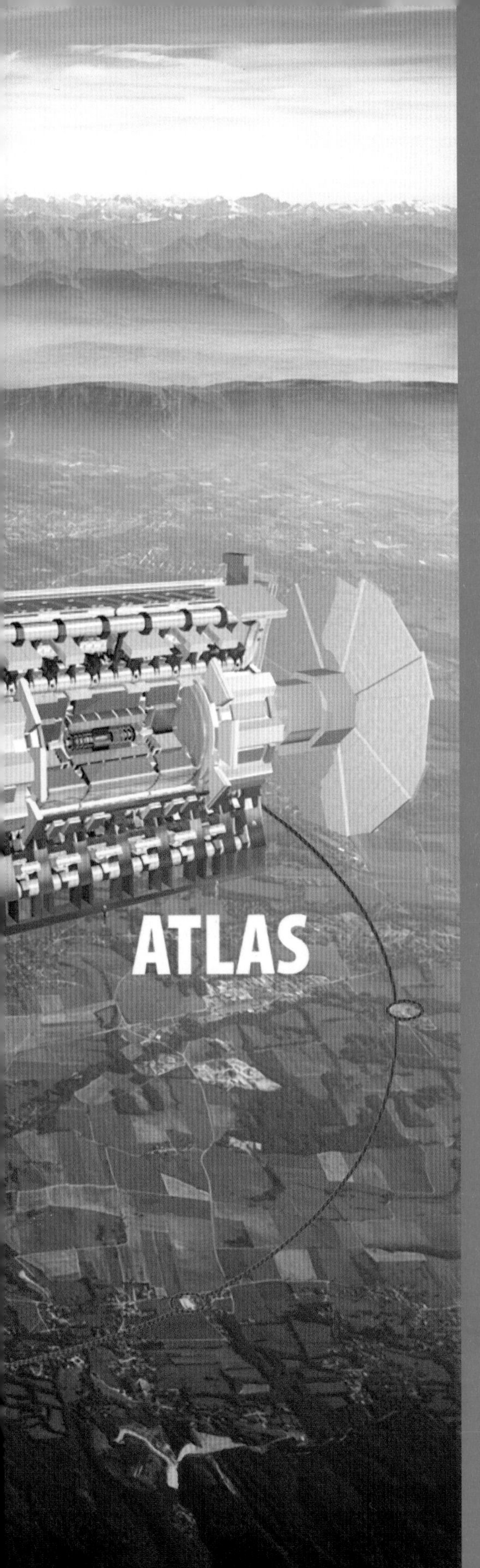

What is a hadron anyway?

The Large Hadron Collider is, quite simply, the latest and most powerful in a long line of particle accelerators.

Inside this circular accelerator, which has a circumference of 27 kilometres, a set of sub-atomic particles racing clockwise at high speed crashes into a set of identical particles speeding anticlockwise. The energy involved in the collisions is so great that, in a sub-microscopic region at the heart of the collision, it briefly generates conditions similar to those that occurred shortly after the birth of the Universe.

Brand new heavy particles flicker fleetingly into existence under the high-energy conditions. This is because Einstein's famous formula, $E=mc^2$ tells us that energy (E) and mass (m) are related to one another via a constant, the speed of light (c). The high energies reached in the LHC make it possible to create particles too heavy to have been produced at lower energy colliders – particles that help physicists build a more complete picture of the subatomic world.

The collisions within the LHC usually involve protons, a type of subatomic particle found in the nucleus of every atom. The LHC owes its name to the fact that protons belong to a class of particles called hadrons.

The two beams of protons, or hadrons, collide at just four points around the circular accelerator. One of four large detectors surrounds each of these collision points, analysing the particles created in the collisions. Of the four, ATLAS and CMS are the largest and most complex.

◀ The LHC ring, with Lake Geneva and the Alps visible behind. The ATLAS and CMS detectors (not to scale) intersect the ring.

2388

◂
A section of the circular LHC tunnel. One of the magnets that steer the subatomic particles around the collider has been virtually opened (through the addition of an overlay) to show the two particle beams inside.

Forty-six metres long, 25 metres tall

Why is the ATLAS detector, which is designed to detect some of the smallest things in the Universe, the length and height of the largest dinosaurs that roamed the Earth? There's an explosive explanation.

Physicists first began making plans to build the ATLAS detector in 1989, a few years after the initial ideas for building the LHC were put on the table. One of the first things they did was to draw up a wish list of particles they thought the new detector might see. It seemed a safe bet that the famous Higgs boson would make its long-awaited first appearance in a physics experiment, but there were also hopes for a raft of other new finds.

The Higgs boson and these other putative particles are so unstable that a few moments after they form they decay again into lighter, stable particles. These lighter particles shoot away from one another in all directions, making a decaying Higgs boson a bit like a subatomic firework explosion. To find the Higgs boson and other new particles, the ATLAS detector would need to carefully track and identify the components of these tiny explosions. Like forensic scientists, physicists then use this information to establish the identity of the original unstable particle.

The ATLAS team began to run computer simulations to work out what, in theory, a decaying Higgs boson should look like – and how complex and sophisticated a detector they would need to give them their best chance of identifying its decay products.

The more simulations the ATLAS team ran, the more they realised that their best chance of identifying new particles including the Higgs boson would come from building a detector with several layers that together created an instrument about 46 metres long and 25 metres tall. The detector's size would allow them not only to catch the highly energetic particles produced in each collision event, but also to measure them very precisely – the key to finding out if the collision had generated anything new and exciting.

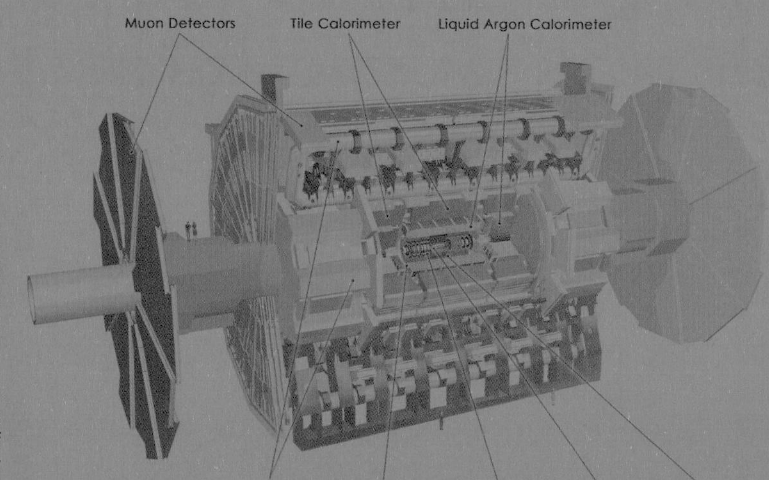

▶
A computer generated image of the ATLAS detector.

▶▶
A technical drawing of the detector gives a sense of its complexity. Assembly took several years.

View from side "C" to IP
Sector 6
Sector 11
Sector 13

A many-layered thing

By the early 1990s the ATLAS detector was shaping up to be one of the largest of its kind ever constructed. What should go inside?

The detector's job is to track and identify the fragments created when unstable heavy particles decay. One proposal was to simply construct an iron ball around the decaying particles large enough to absorb most of the subatomic shrapnel they generated – but in the end a more sophisticated design won out.

Today, the ATLAS detector consists of three large detector systems wrapped concentrically around one another. A 1.2-metre-diameter inner detector tracks the precise path each charged particle takes. Around this is an 8-metre-diameter calorimeter that absorbs most of the particles, and in doing so measures their energy. Finally, around the calorimeter there is a 22-metre-wide muon spectrometer, which measures a special class of particles – the muons – that penetrate material very easily and so plough right through the calorimeter.

The layers of the detector: collisions occurring at the detector's centre (white circle) release a spray of subatomic particles that travel through the inner detector (green), the calorimeter (brown and blue) and finally into the muon spectrometer.

The calorimeter after installation, in April 2008.

Overleaf: Part of the inner detector.

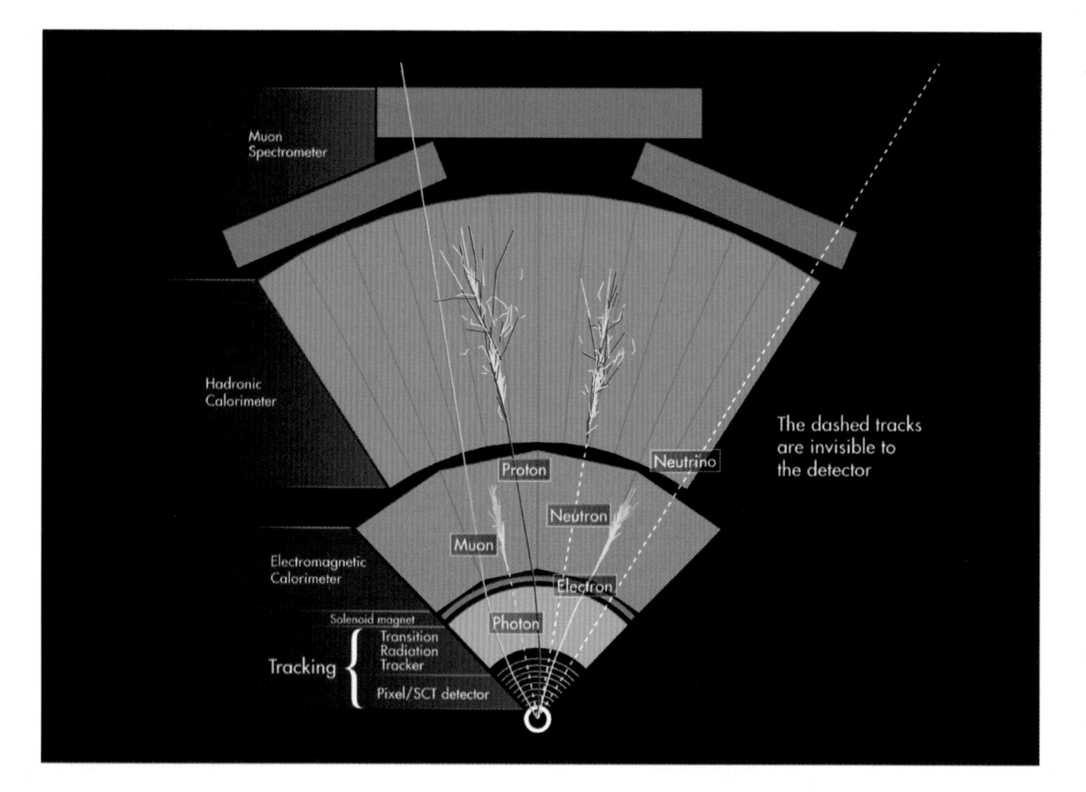

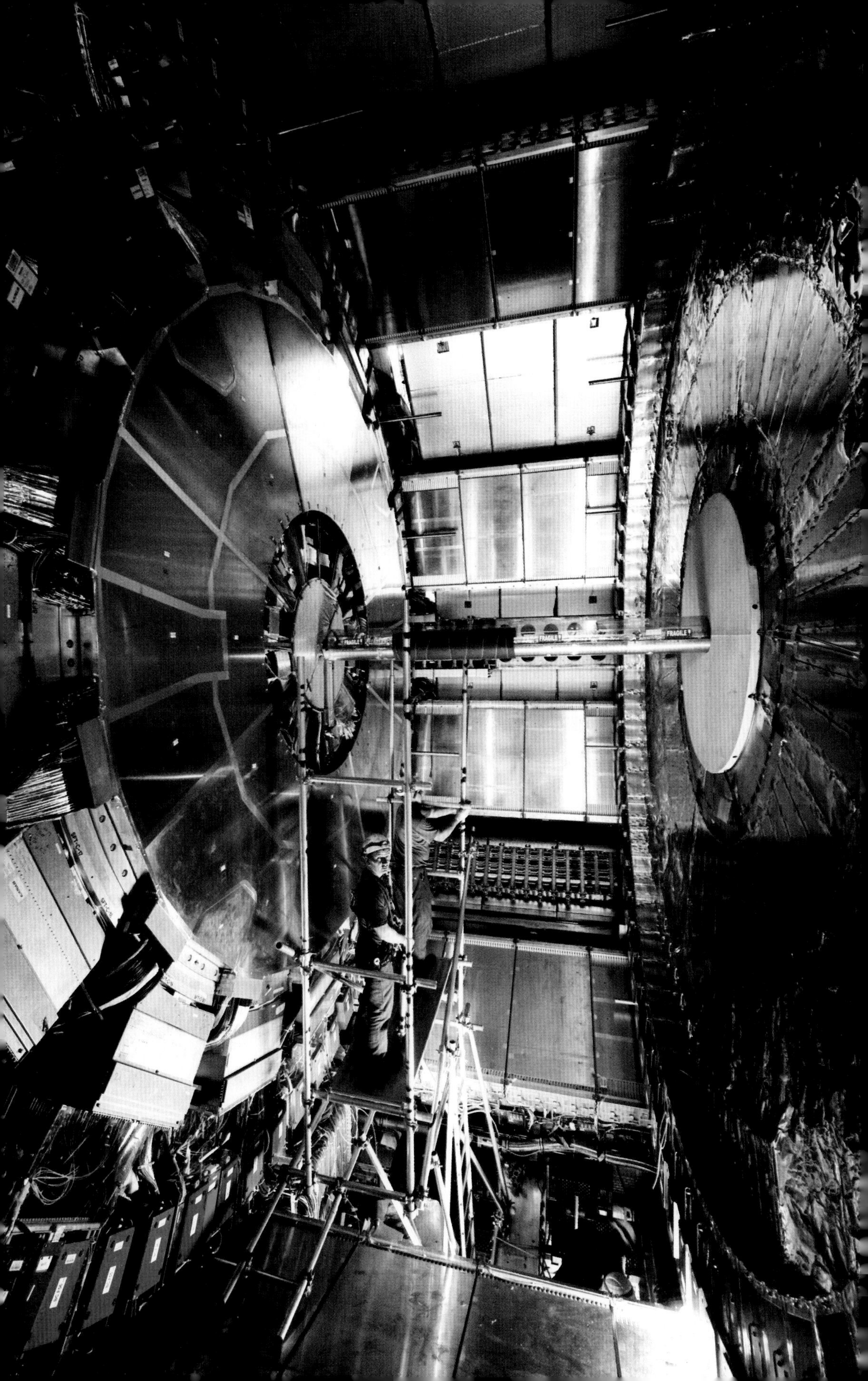
FRAGILE!
FRAGILE!
FRAGILE!

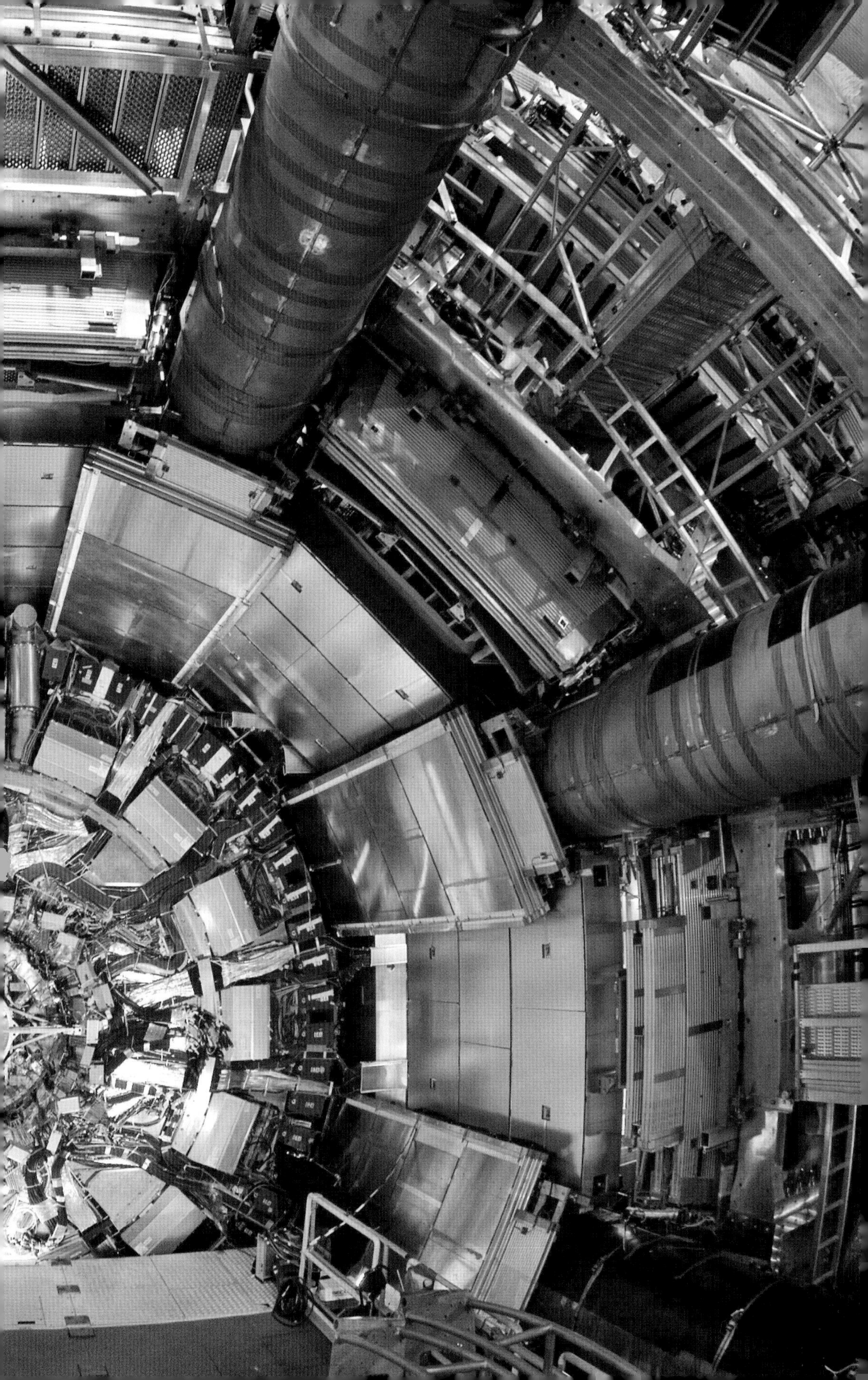

Magnets within the detector bend the electrically charged particles in a way that makes it possible to measure some of their properties. From the outset, some physicists proposed building these giant magnets out of superconducting materials. Superconductors are a special class of material that, when cooled to temperatures close to absolute zero (a few hundred degrees below the freezing point of water) allow electricity to flow through them without any resistance. They make very strong magnets. But in the early 1990s no one had ever built superconducting magnets as large as those proposed for the ATLAS detector – they were a truly ambitious design. By the time the ATLAS detector was switched on 18 years later they had become reality.

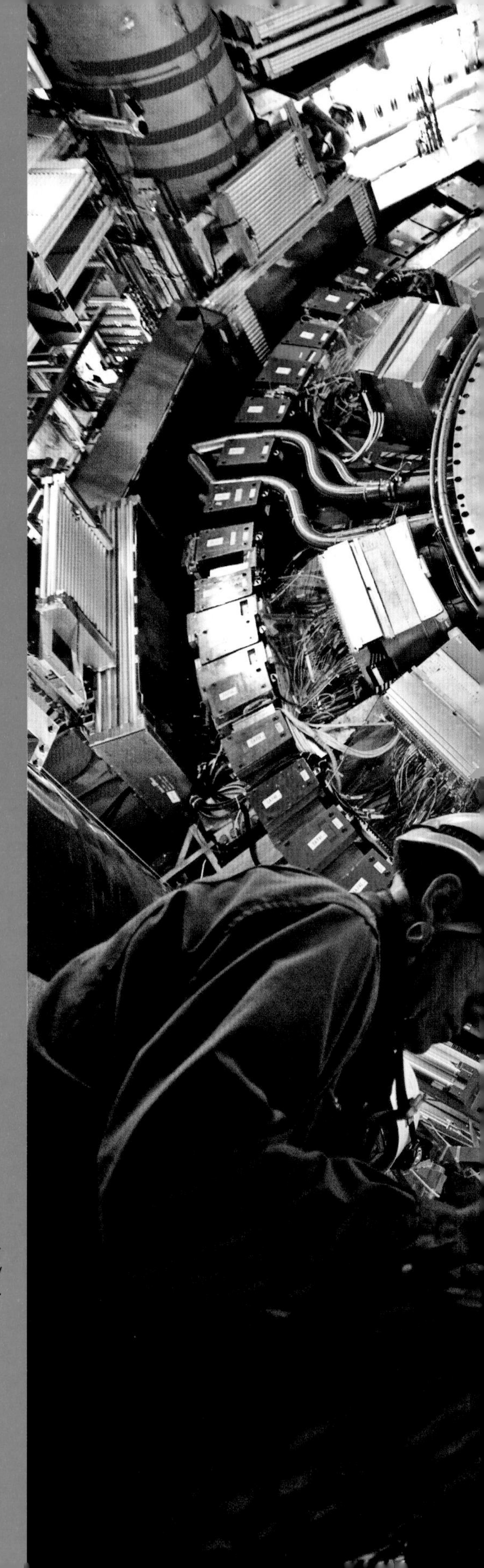

The calorimeter on the move in January 2011 during a maintenance period.

Overleaf: Big wheels of detectors on either side of ATLAS form part of the muon spectrometer.

Page 24-25: Eight giant superconducting magnet coils stretch backwards towards the circular calorimeter during installation.

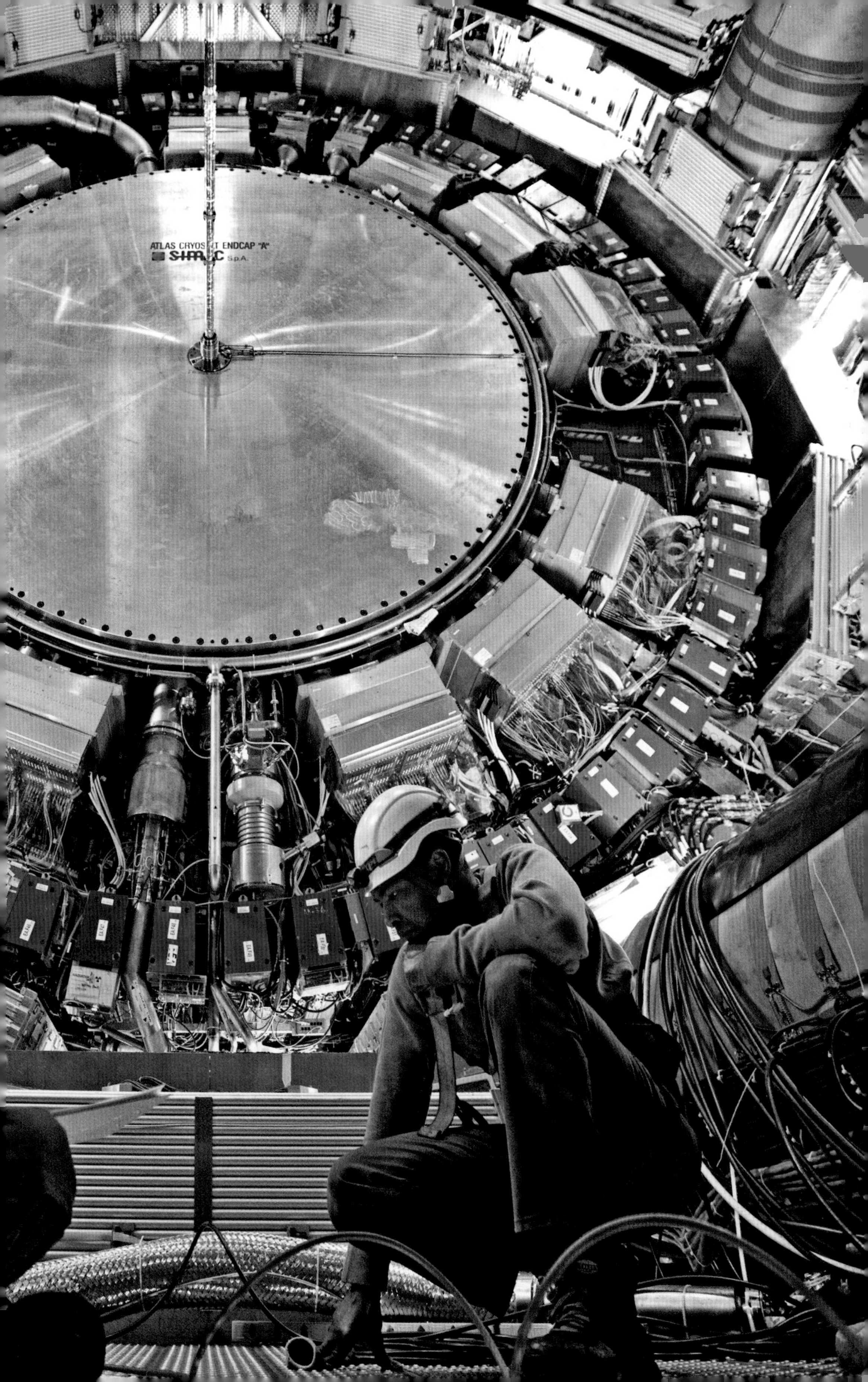
ATLAS CRYOS T ENDCAP "A"
S.p.A.

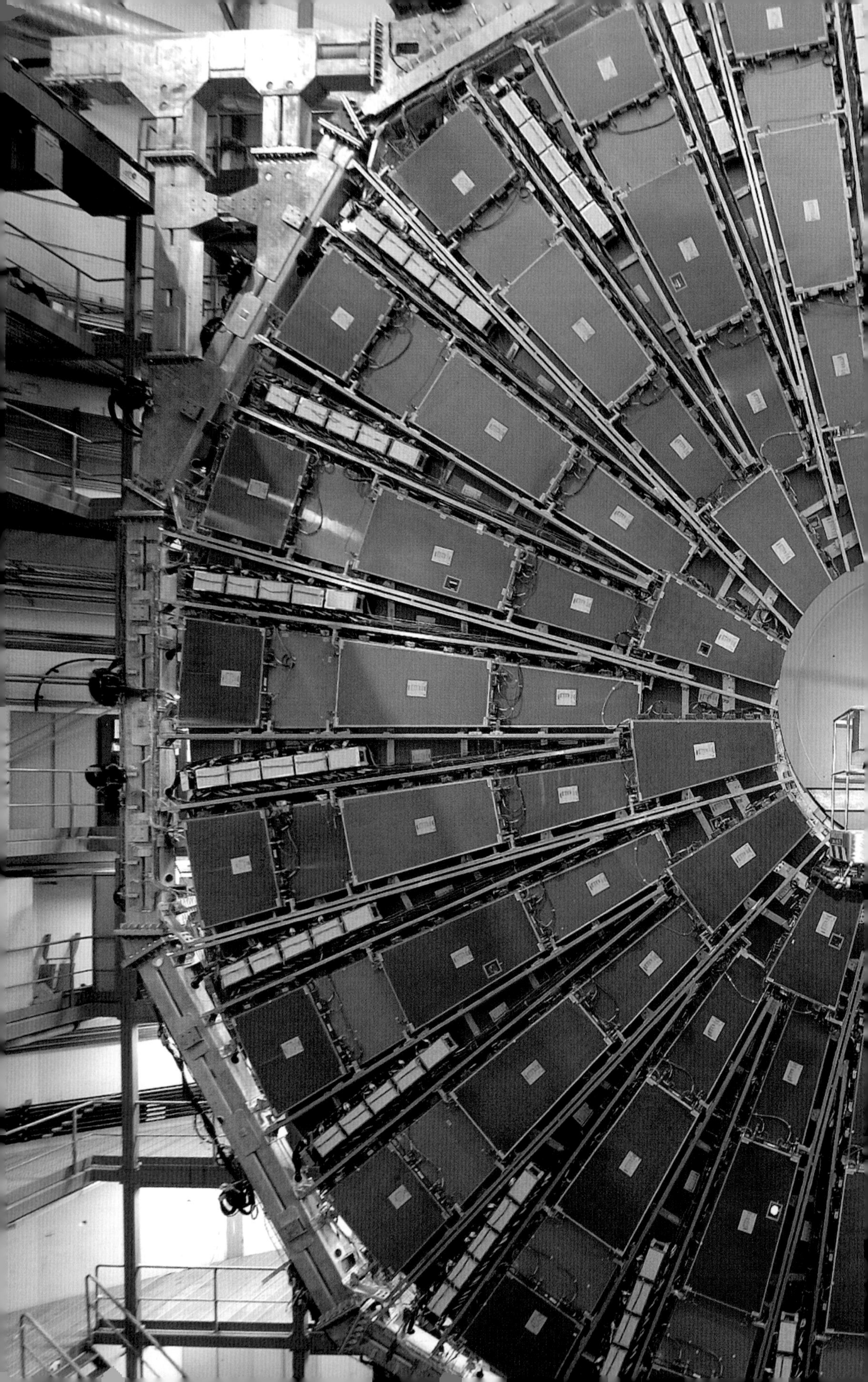

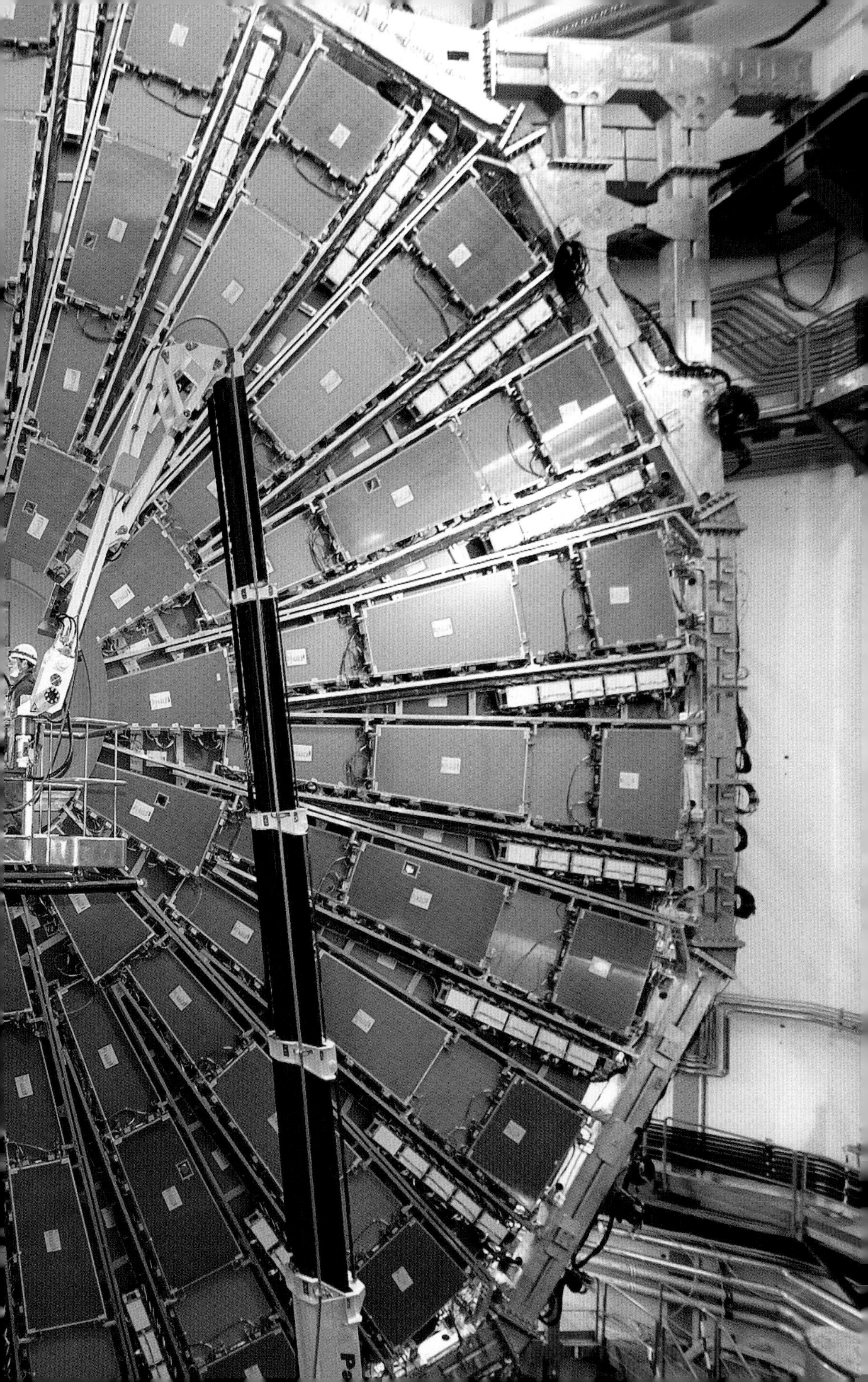

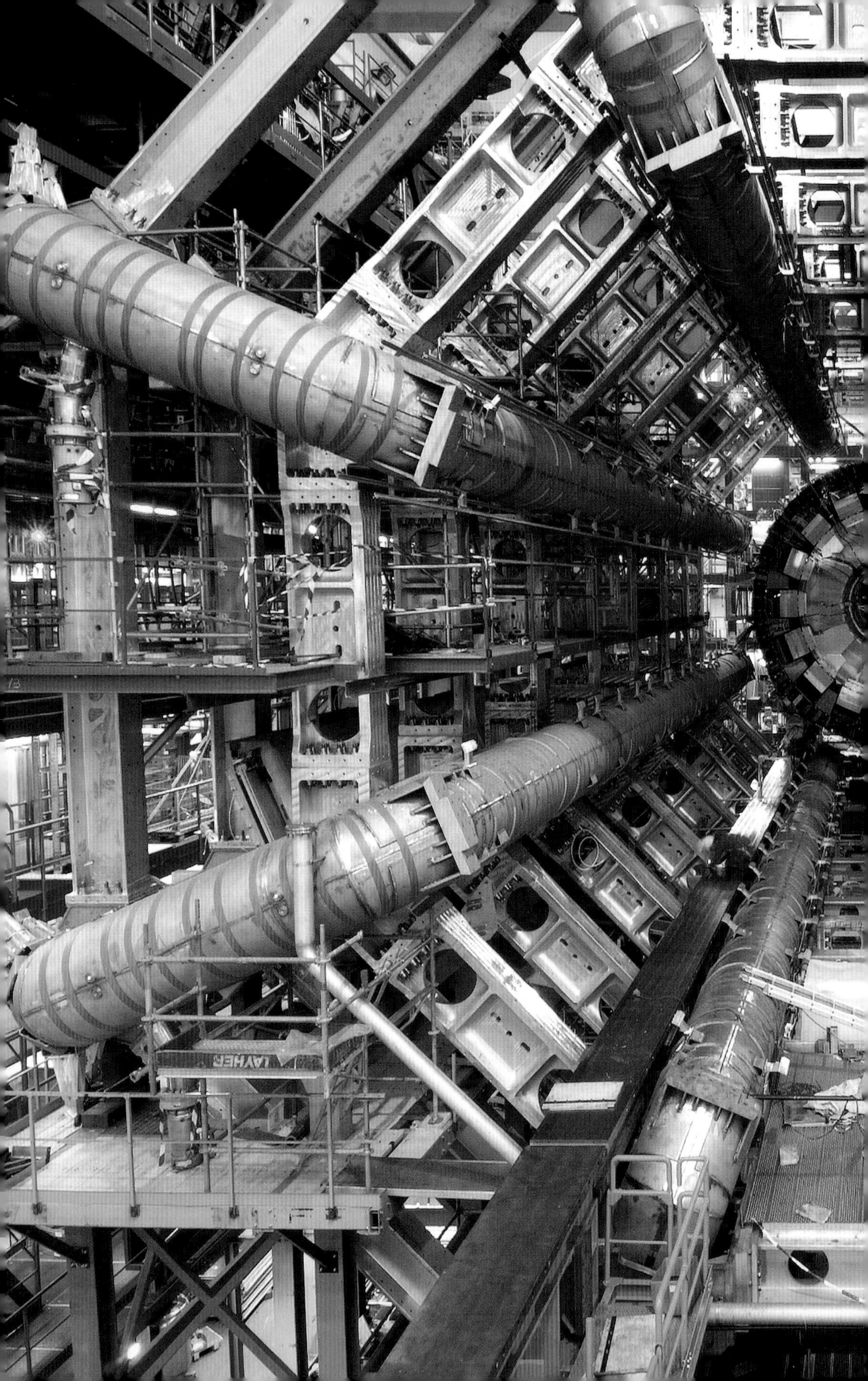

From the four corners of the world

> "Bits and pieces of the ATLAS detector were built the world over, and then transported to Geneva for assembly"

As plans to build the ATLAS detector grew, so did the size of the ATLAS team of physicists and engineers. By the mid 1990s, around 150 institutions in 32 countries were on board. Building the world's largest detector would be a collaborative effort involving them all.

Bits and pieces of the ATLAS detector were built in labs and factories the world over, and then transported to Geneva for assembly. Some of these components were small enough for one person to carry; others so large that transport was possible only with the aid of lorries, barges, trains and planes. By 1998, the first pieces had begun to arrive at CERN for pre-assembly and by June 2003 the installation work could begin in the cavern, 100 metres below ground.

Inner detector

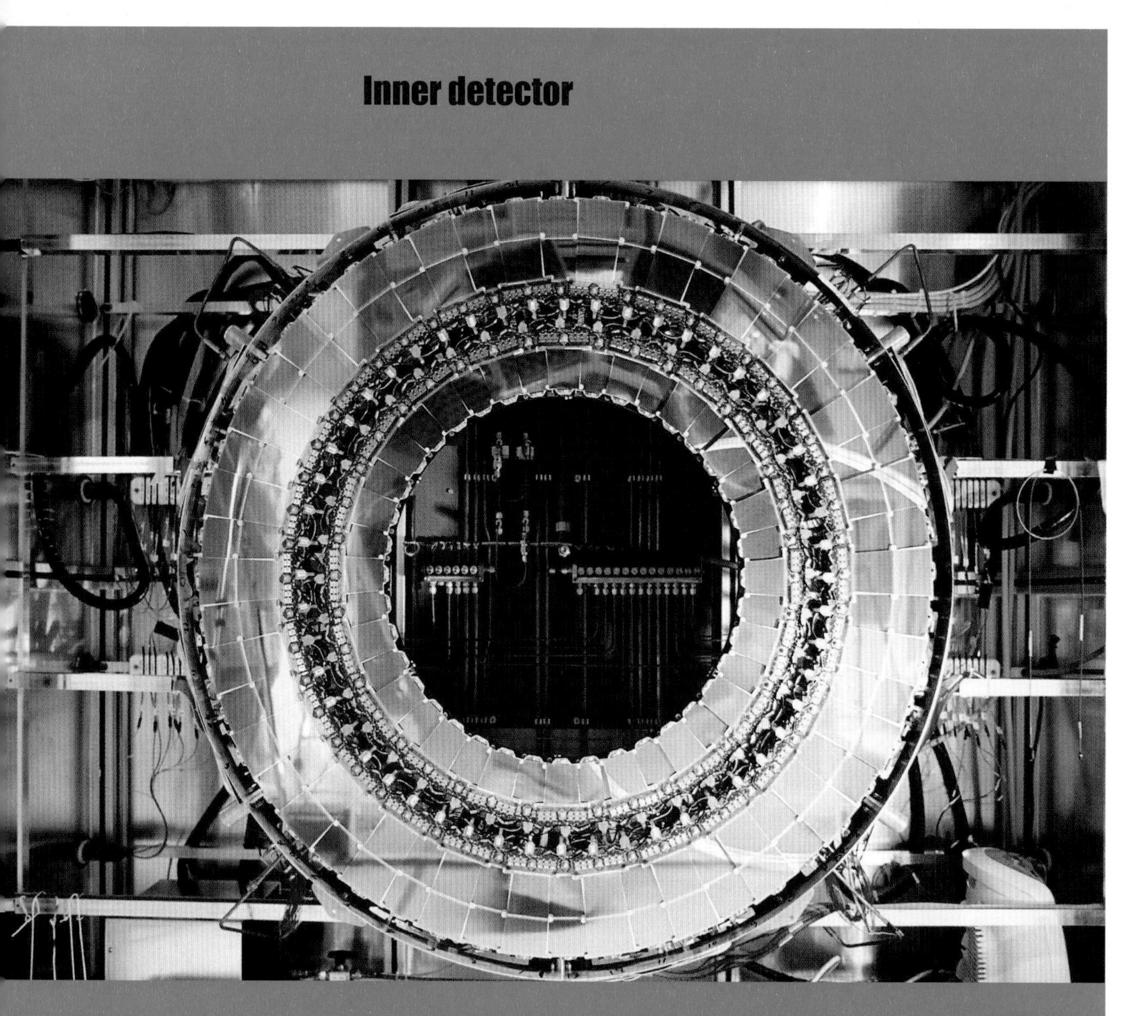

Calorimeter construction

▸

Half a million pieces of blue scintillation plastic, half a million plates of steel and 187,000 pieces of green fibre make up the multi-layered tile calorimeter.

The 4000 individual muon chambers were built using four different technologies by 48 institutions in 23 production sites around the world.

▾

◂

The inner detector is the most compact subsystem in ATLAS and it needs to be highly sensitive: it is the first part of ATLAS to see the decay products of the collisions.

Muon chamber construction

Putting it all together

With a volume of 50,000 cubic metres – equivalent to about half the size of the Notre Dame Cathedral in Paris – the ATLAS detector cavern was a truly impressive sight even when still empty. Almost all the available space is now filled by the particle detector.

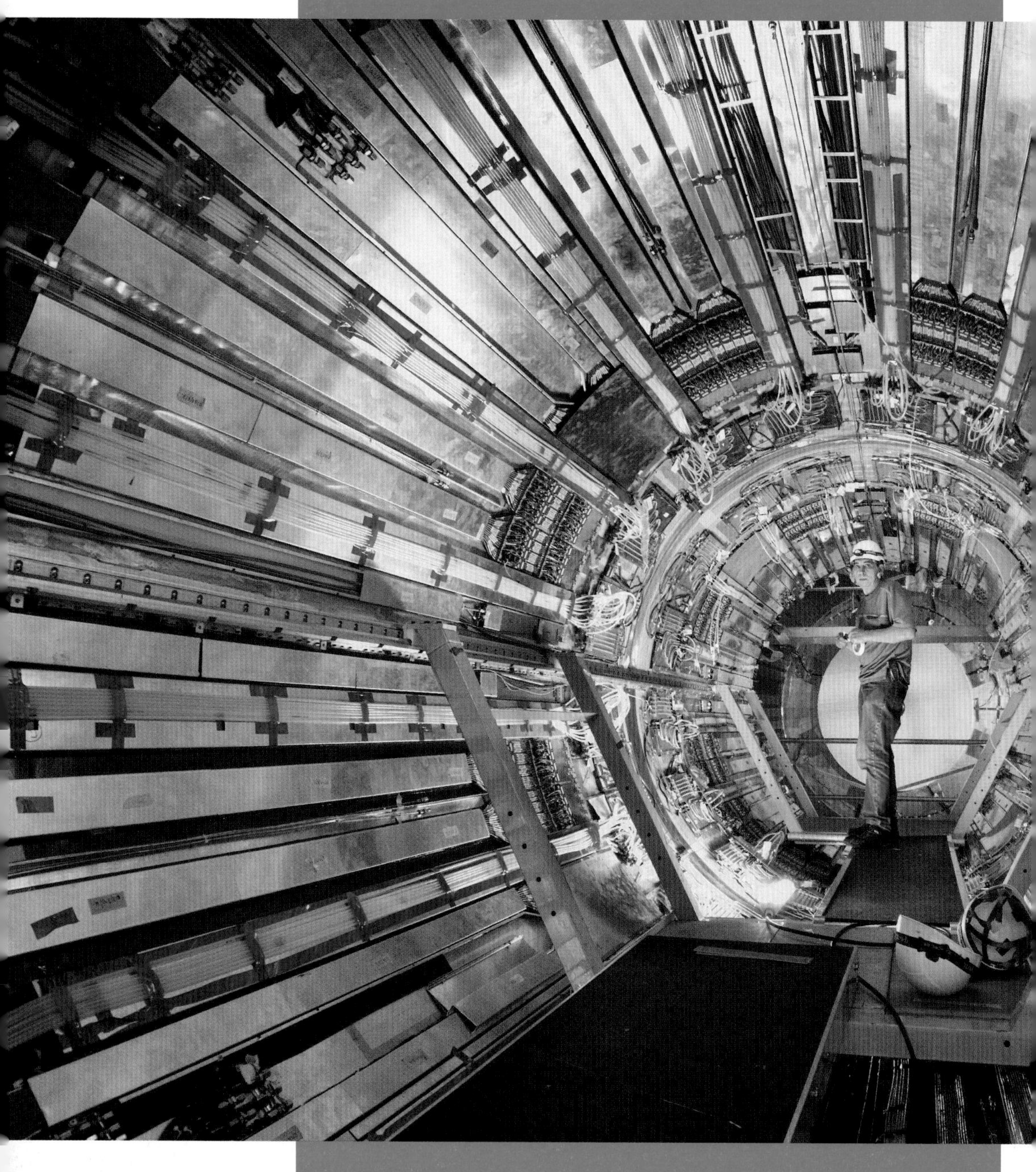

▸ Working on one side of the ATLAS cavern during the last week of January 2008.

A scientist works on the inner wall of the cryostat. During operation, it contains liquid argon at -183 °C to keep the calorimeter cold.
▾

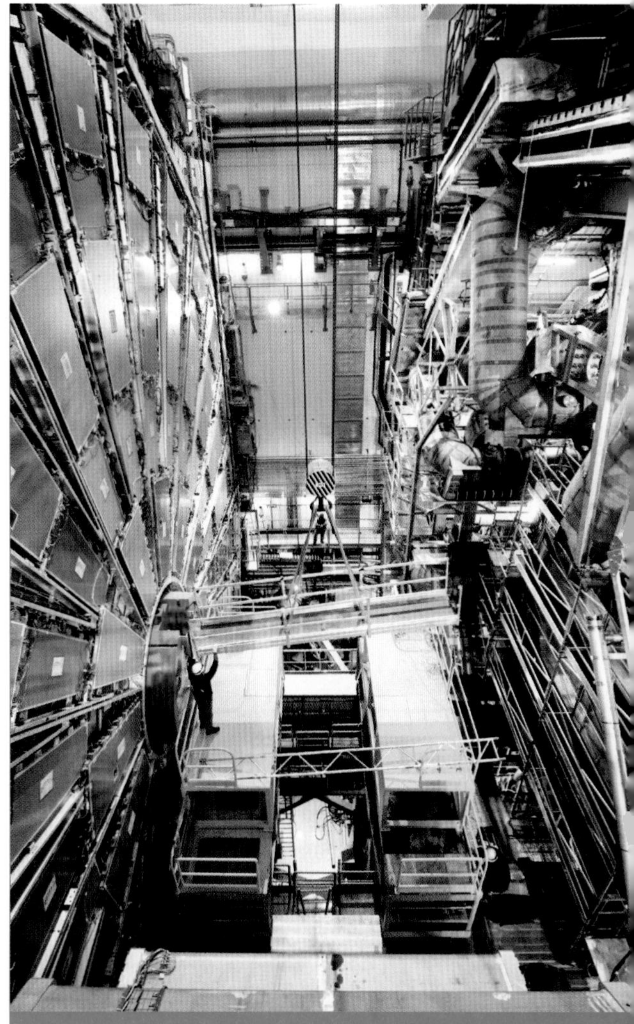

Assembling the ATLAS detector in this cavern has been likened to building a ship in a bottle. The description is apt, since the large detector pieces had to be lowered into the cavern down one of two vertical "bottlenecks" – access shafts just 18 metres and 14.5 metres in diameter, respectively. The largest and heaviest detector components fit with just millimetres of clearance to spare on each side.

One by one, the various pieces of the ATLAS detector descended into the cavern. It would take around 60 months to fit them together, but by May 2008 all the pieces were in place.

▲ The innermost part of the inner detector, which is known as the pixel detector, during construction in the lab.

▲ The inner detector's silicon trackers had to be built in very clean conditions.

◀ A component of the inner detector ready to slot into position.

Overleaf: A section of the calorimeter following insertion into the cryostat that keeps it cool when the detector is operational.

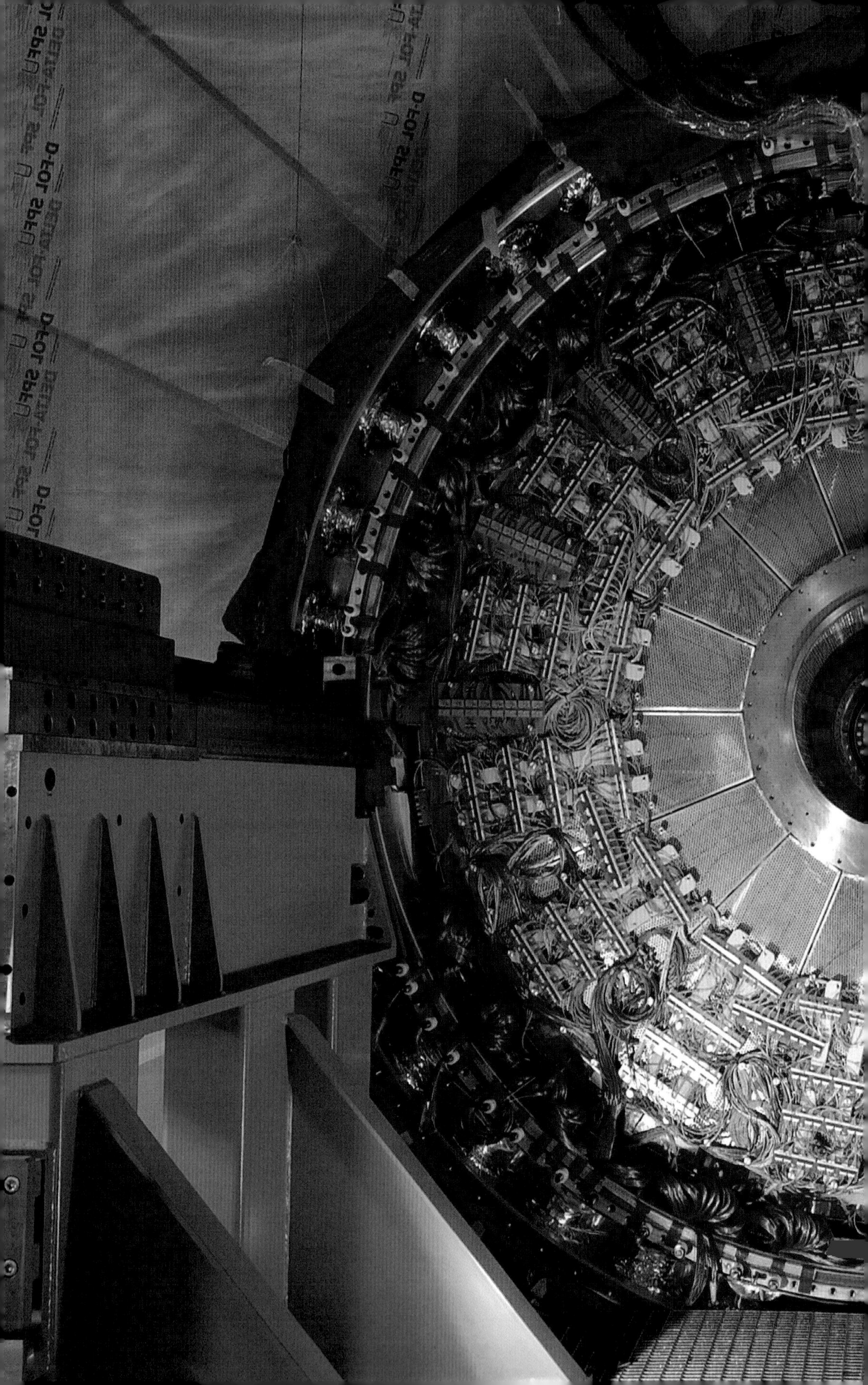

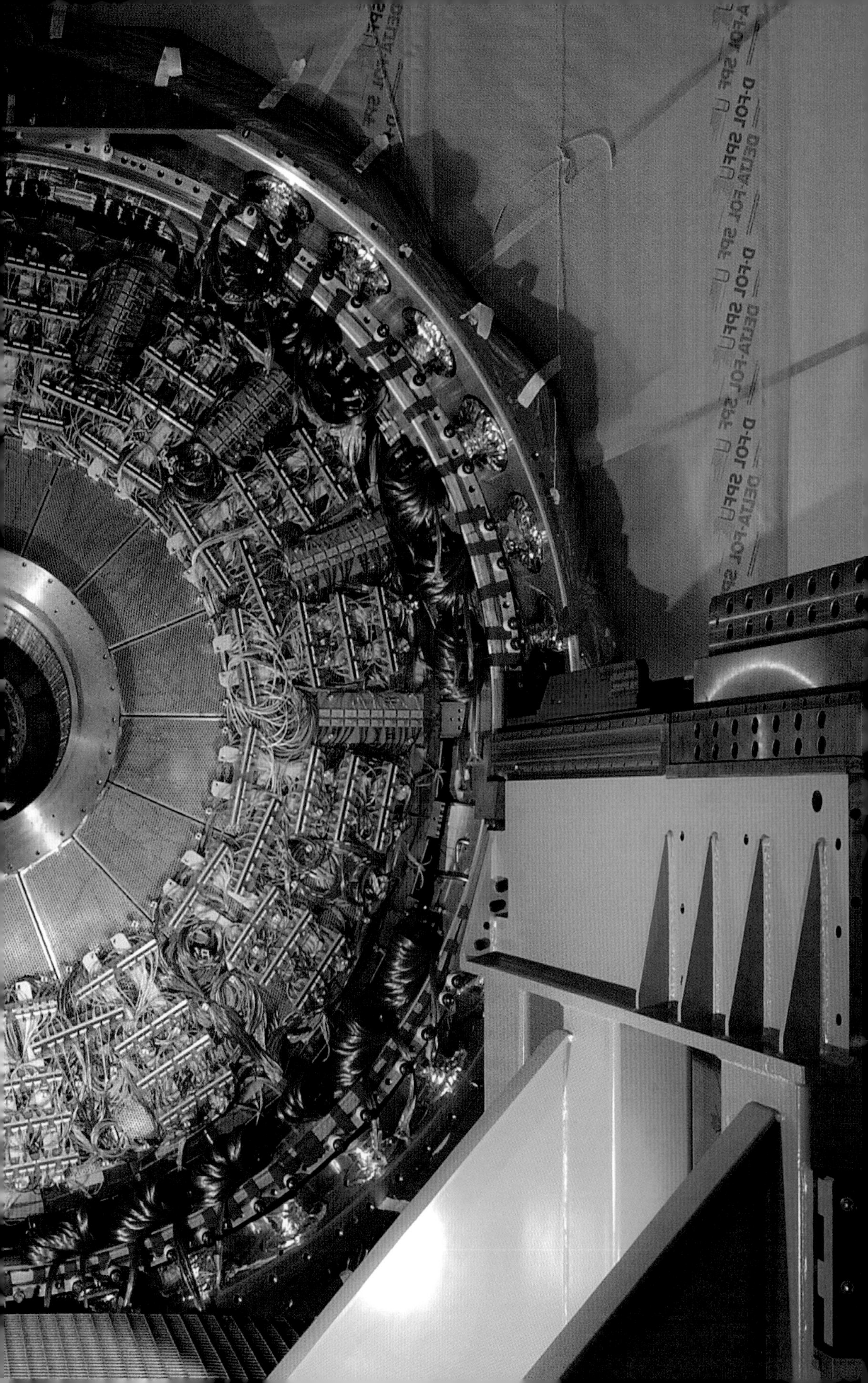
D-FOL SPF
DELTA-FOL SPF

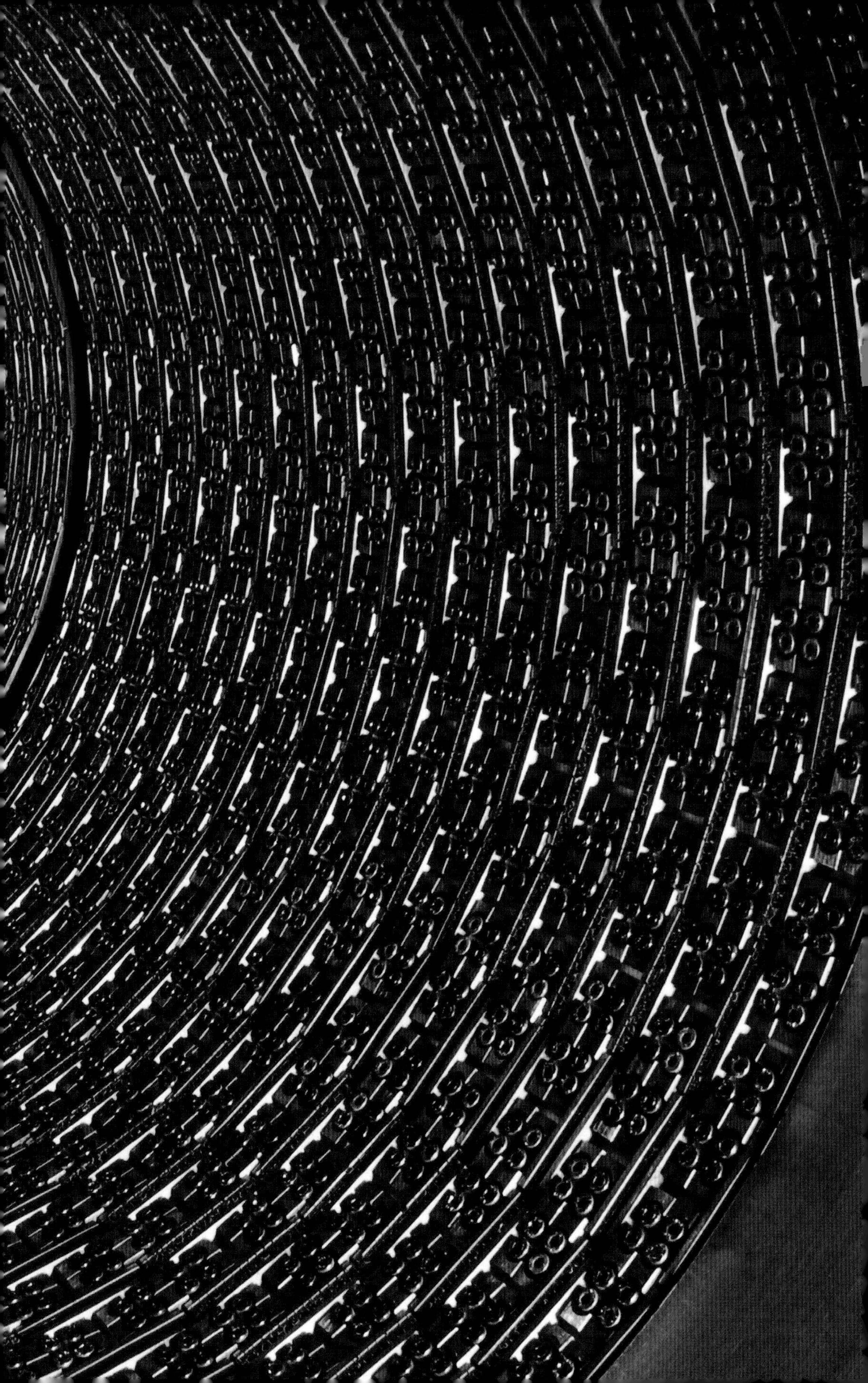

Previous page: A close up of the ATLAS hadronic endcap liquid argon calorimeter.

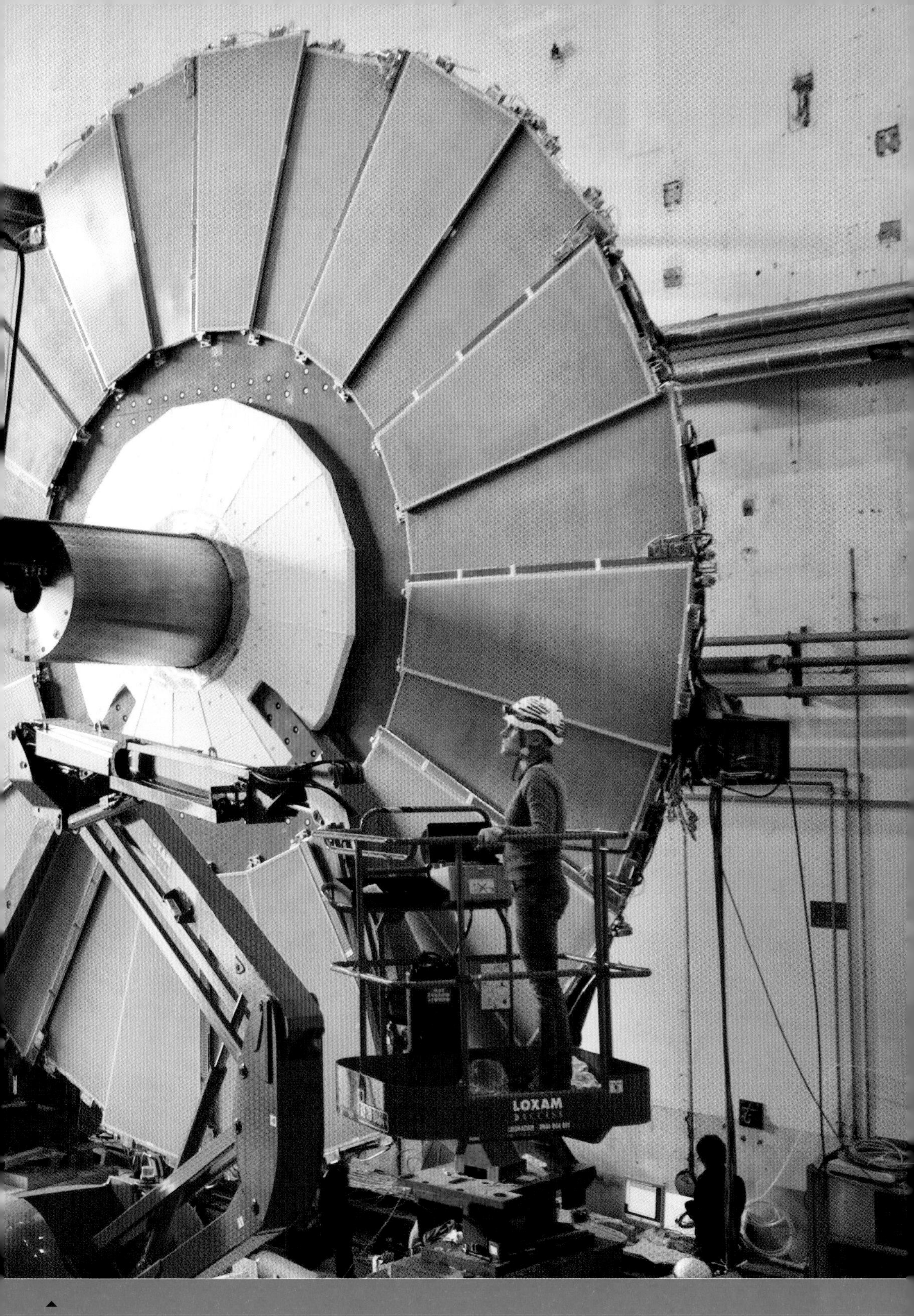

▲
One of the surprisingly large "small wheels" – part of the muon spectrometer – during construction.

> **When cooled to temperatures close to absolute zero, super-conductors allow electricity to flow through them without any resistance. They make very strong magnets**

◄ One of the end caps – which includes part of the cooling system for the ATLAS detector's huge superconducting magnets – and some members of the team that built it.

◀▲ The first of eight superconducting magnet coils was lowered into the ATLAS cavern on 26 October 2004.

Turn on, tune in

ATLAS is one of the most ambitious physics experiments ever run. Before the ATLAS team could begin using their detector to make new discoveries, they had to check that it was functioning properly – and that they could make sense of what it told them.

Even during the 5-year-long installation process, the ATLAS detector was subject to a rigorous regime of tests. The LHC itself was not ready to provide the growing detector with a supply of subatomic particles to study, so the ATLAS team looked to the heavens for inspiration.

Cosmic rays are very high-energy particles that originate in space. They collide with particles high in Earth's atmosphere, and some of the new particles generated in these collisions rain down on Earth with enough energy to penetrate our planet's surface. The ATLAS detector may be 100 metres below ground, but it is naturally bathed in a shower of subatomic particles called muons generated in collisions thousands of metres above our heads.

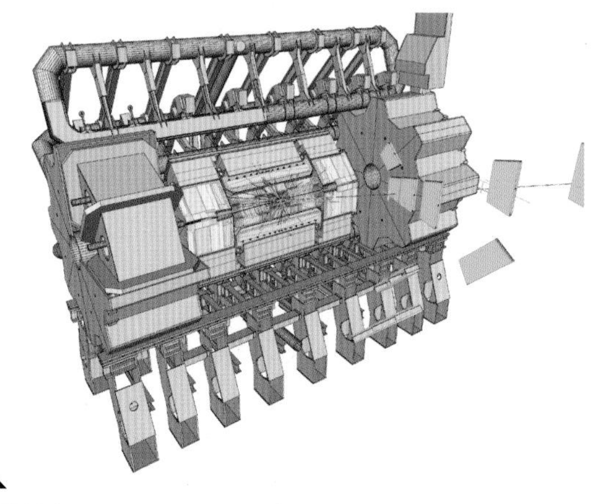

ATLAS detector cut-away view showing a simulated collision.

The ATLAS team began using their growing detector to track these high-energy particles, offering them an opportunity to iron out any problems long before the programme of physics discovery began.

In November 2009, the LHC began smashing protons together inside the ATLAS detector. With real data from these proton-proton collisions, the ATLAS team began a four-month period of final checks and calibrations. Turned on and now tuned in, the ATLAS detector was ready to explore the subatomic world for new physics.

Reaction to the first collisions, in November 2009.

A computer display shows some of the first collisions reconstructed within the inner detector, on 6 December 2009.

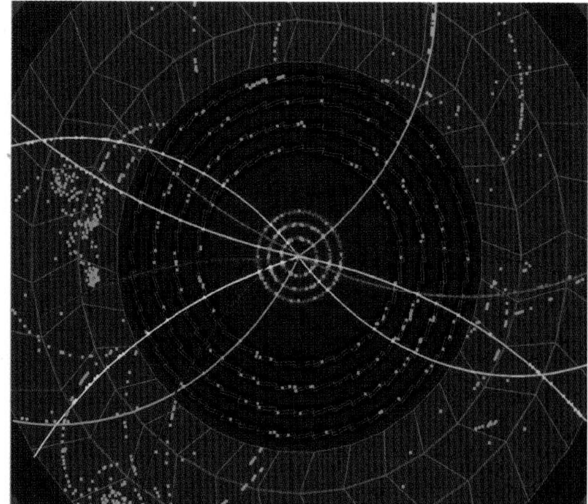

Extraordinary drops in an ordinary ocean

In March 2010, the LHC began to collide protons together at energies never before achieved in a physics experiment. The ATLAS detector could get on with the two tasks it was designed to perform: measure the familiar with far greater precision than achieved before, and look out for the unusual. The second task is particularly difficult, though: every second, several hundred million pairs of protons smash together deep inside the ATLAS detector – far too many collisions to record and store. Only a tiny fraction can be kept for detailed study. The ATLAS team needs a way to quickly sift through the haystack in search of the needles – the tiny number of unusual decay patterns.

The only option here is to allow the detector itself to do the bulk of the sifting. Although its primary task is to record the collisions occurring at its centre, the ATLAS detector's electronics double as a giant sieve. They have been built and trained to recognise and reject the data from the less interesting collisions in the blink of an eye. And they are very discerning: in just a fraction of a second they automatically identify the one collision in every 10,000 that is likely to interest physicists and save only these events for further analysis.

Even this is too much to store. Instead, the ATLAS detector electronics pass the collision events that survived the first selection process to a large computing farm located above the detector cavern, on the surface. Computer software sifts through the data twice more to find the roughly 400 most interesting proton-pair collisions that occur each second. These 400 are equivalent to only 0.002% of all beam crossings happening inside ATLAS. They add up to a more manageable 600 megabytes and would fit on one CD. It is this CD's-worth of information generated each second that actually leaves the experimental site for the ATLAS team around the world to study.

One of the many tunnels around the ATLAS cavern that bring the cabling and supply pipes to the ATLAS detector.

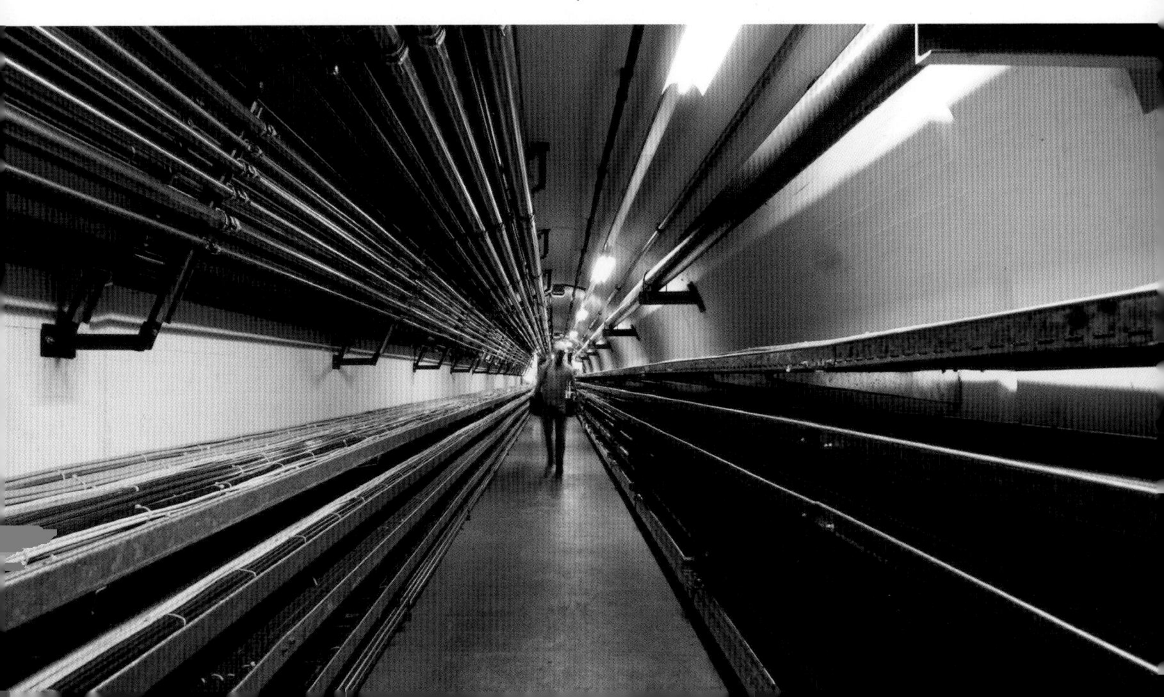

Working round-the-clock – and around the world

The ATLAS detector automatically discards 99.998% of the collision data it generates. But working with the remaining 0.002% is no easy task. Between 2010 and 2012, this 0.002% added up to about 125 million gigabytes. To study this information both thoroughly and quickly, the thousands of ATLAS physicists based the world over need fast and repeated access to the stored data.

The internet is a convenient tool for sending information quickly from country to country. However, the volume of data generated by the ATLAS detector is too large for the internet to comfortably handle at the same time that it deals with traffic generated by the 2 billion people who are regularly online. CERN designed and built an alternative system specifically to distribute data from ATLAS and the other LHC experiments. This is called the Worldwide LHC Computing Grid.

▲ Some of the electronics of the ATLAS trigger system, which selects the data that will be kept for further analysis.

▲ Data storage within the CERN computer centre.

The internet has no central point, but the Worldwide LHC Computing Grid does. It is the CERN computing centre, which is where the information from the ATLAS detector is sent first. After some initial processing, copies of the data leave this central point and head to eleven large hubs in Europe, North America and Asia. These hubs then make the data available to the (at the time of writing) 177 institutes in 38 countries that participate in the ATLAS collaboration. Now the 3000 or so ATLAS physicists can access the data at their institutes or at CERN, to pursue their research goals. Just days after a collision occurs inside the ATLAS detector, it may be studied by a physicist thousands of miles from Geneva.

▲
Inside the CERN computer centre.

In June 2012, the ATLAS team began working even faster to process data so that they could present up-to-the-minute conclusions at an important international physics conference taking place early in July. Working around the world and round-the-clock, physicists discussed their results with increasing excitement as the deadline approached. Finally the ATLAS team, and their counterparts in the CMS experiment, were ready to make a very significant announcement.

The final year of the Higgs Hunt

> **"By December 2011, both the ATLAS and CMS teams were seeing hints of an excessive peak in activity around 125 GeV"**

In the 1960s, several physicists proposed a mechanism to assign mass to particles, which predicted the existence of the Higgs boson. The particle has been evading detection ever since, though, and so no one expected it to instantly pop into view inside the ATLAS detector. In fact, the ATLAS team thought that only once they had analysed several hundred trillion collisions could they hope to identify the Higgs boson's subtle signal.

By June 2011, ATLAS and its sister detector, CMS, had together accumulated data from about 70 trillion collisions – a figure that physicists describe as 1 inverse femtobarn. Within the data were signs of something unusual in the mass range 120-150 gigaelectronvolts, or GeV (remember that energy and mass are closely related). This is relatively heavy in the subatomic world – about 125 times heavier than a hydrogen atom. Speculation began that these signals might contain a new, comparatively heavy particle. Was this the missing Higgs boson? More data would hold the answer.

By December 2011, the ATLAS and CMS teams had gathered 5 inverse femtobarns of data. Both detectors were now showing hints of a new particle around 125 GeV, but still the signal was too weak to be sure. However the signal strengthened when more collisions were analysed, and by July 2012 both the ATLAS and CMS detector teams could report with confidence that a new particle had been discovered.

▼ The matter particles of the Standard Model of particle physics.

▶ Peter Higgs at CERN in April 2008.

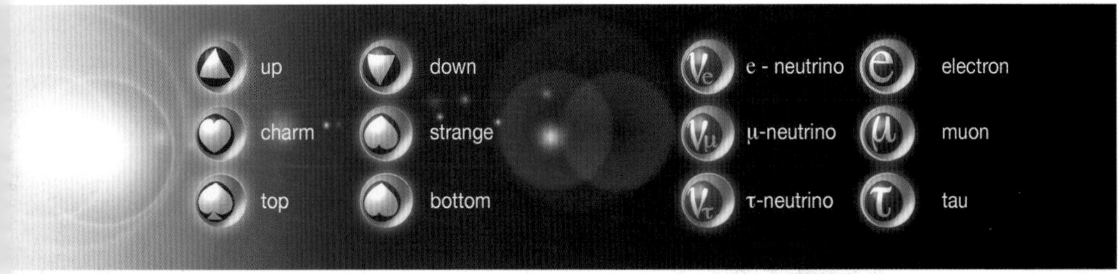

up
down
e - neutrino
electron
charm
strange
μ-neutrino
muon
top
bottom
τ-neutrino
tau

"We have reached a milestone in our understanding of nature"

Rolf Heuer, CERN Director General

▲ The ATLAS Experiment spokesperson Fabiola Gianotti announces the discovery of a new particle at a joint seminar with the CMS Experiment at CERN.

▶ A Higgs boson may decay into two pairs of electrons, as seen here by the ATLAS detector. The electron pair tracks are coloured red and blue.

On 4 July 2012, just 2 years and three months after the LHC first began delivering proton-proton collisions at energies higher than any other experiment has achieved, the ATLAS and CMS teams had recorded enough data to identify a new particle. This new particle shows every sign of being the Higgs boson. Why is its discovery such a milestone?

During the second half of the twentieth century, physicists pieced together a theory of the subatomic world that explains the very nature of matter – describing it in terms of twelve fundamental matter particles (and their antiparticles) interacting via three fundamental forces. There is a fourth fundamental force – gravity – but the individual particles are much too light to feel any significant gravitational effect.

This Standard Model of particle physics has formed the backdrop to a large chunk of physics research over the last 35 years: experiments have been designed and built specifically to identify new subatomic particles that the Standard Model has predicted exist.

ATLAS
EXPERIMENT
http://atlas.ch
Run: 203602
Event: 82614360
Date: 2012-05-18
Time: 29:28:11 CEST

ATLAS
EXPERIMENT
http://atlas.ch

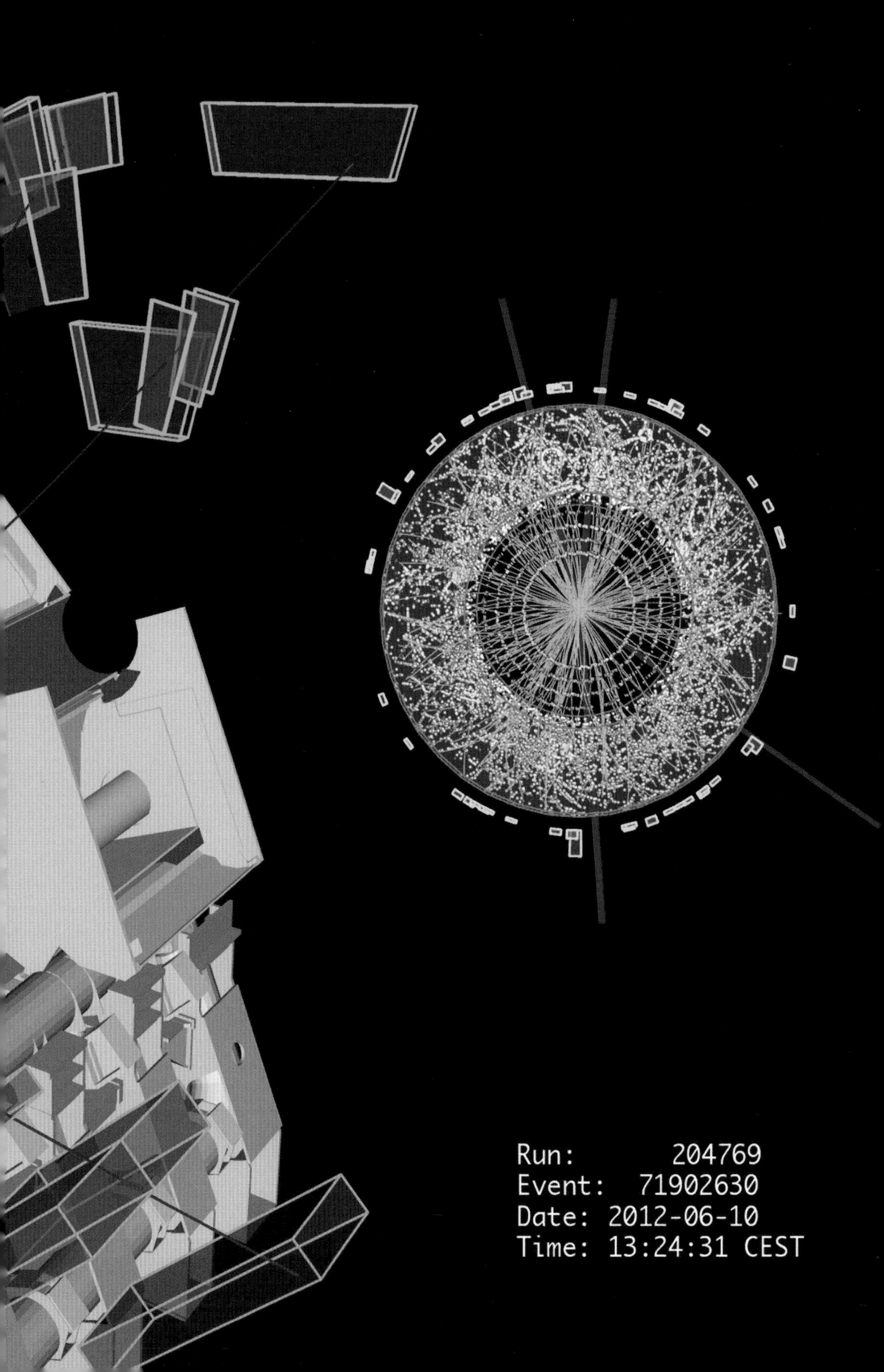
Run: 204769
Event: 71902630
Date: 2012-06-10
Time: 13:24:31 CEST

> Time and again, these experiments have validated the predictions made by the Standard Model

Time and again, these experiments have validated the predictions made by the Standard Model. The most significant success stories include the discoveries, in 1983 and 1995 respectively, of the W and Z bosons and of the top quark. Both discoveries fit neatly with the Standard Model's predictions. But one piece consistently remained elusive in the pre-LHC experiments: the Higgs boson.

Previous spread:
A Higgs boson may also decay into four muons, as observed by the ATLAS detector here. The paths of the four muons through the detector are traced in red.

▲
Experimental results from the ATLAS Higgs search identified a clear peak (left) between 123 and 130 GeV – the signature of the new particle.

The Higgs boson was a particularly unfortunate missing piece, because it has a crucial role to play. It is part of the mechanism that assigns mass to the 12 fundamental particles in the Standard Model. A universe in which they are all weightless would have very different properties from those we are familiar with. Without the Higgs boson, the theory was not complete.

“The Higgs boson was a particularly unfortunate missing piece from the Standard Model, because it has a crucial role to play”

A new beginning

Confirmation of the Higgs boson's discovery marks the end of a journey that began in the 1960s, when the particle was first suggested to exist. It also marks a new beginning.

There is more to particle physics than simply finding new particles. Like a biologist who finds a new species of animal, physicists want to learn all they can about how a new particle behaves. This information provides a way to test some Standard Model predictions at a deeper level, and tighten up other predictions in a way that can guide the direction of future research.

As an example, go back to CERN before the LHC. The physics lab was then home to another particle collider – the Large Electron Positron Collider, or LEP. The LEP collider was switched on in 1989, six years after the Nobel prize-winning discovery of the W and Z bosons, the carrier particles of the weak

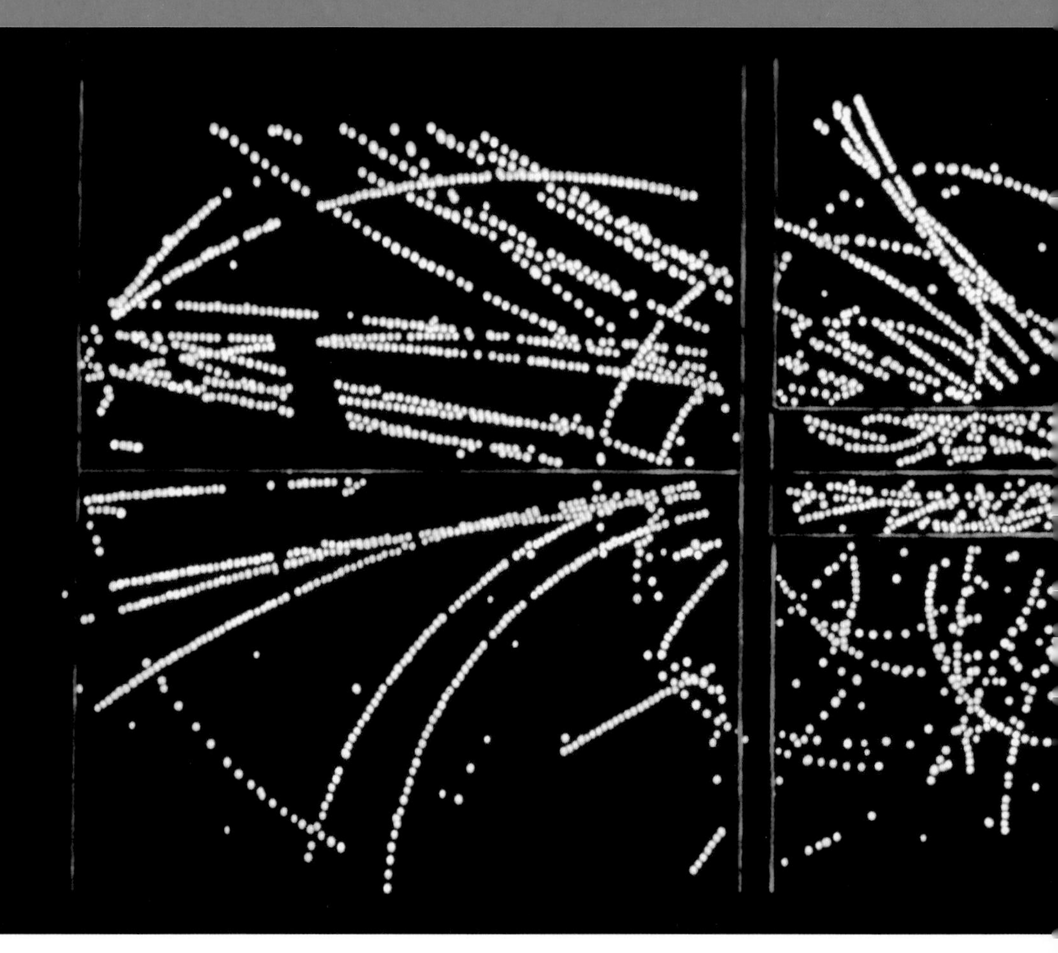

A high-energy electron (arrowed) seen in 1982 by the UA1 detector at CERN. The electron is one of the decay products of a W boson, and helped identify the particle.

force that governs nuclear decay in the Standard Model.

Although the LEP collider did not discover the W and Z bosons, it allowed physicists to measure the properties of both particles with an incredible degree of precision. Doing so revealed tiny shifts in their mass – shifts that made sense if the then still-missing Higgs boson as predicted by the Standard Model was exerting an influence on the W boson. To have the required influence, and given all other measurements of the Standard Model, this Higgs boson would need a mass of no more than 150 GeV.

In other words, long before the LHC began its search for new physics, physicists had strong reasons to suspect that the new collider would discover the Higgs boson roughly where the particle eventually appeared, in July 2012.

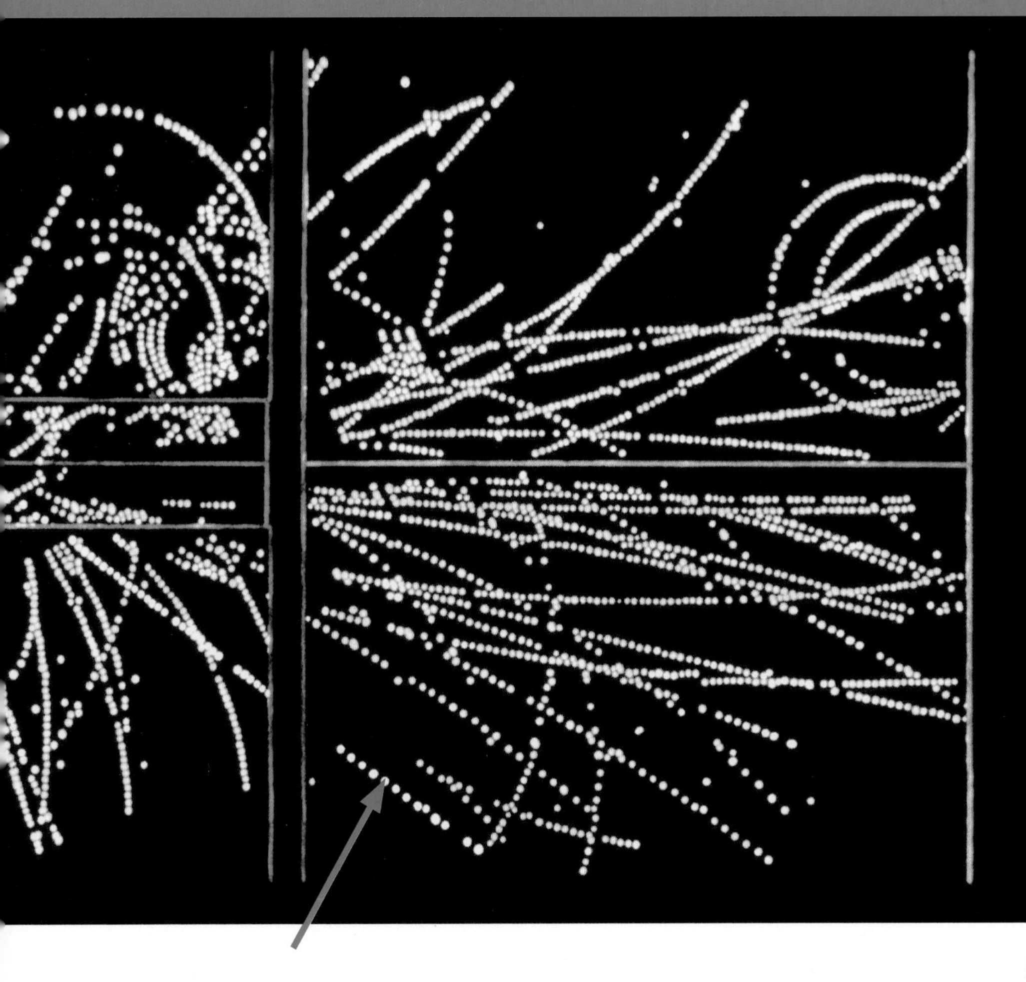

Beyond the Higgs boson

Physicists waited a long time for the LHC, and the ATLAS detector, to begin delivering data. Perhaps surprisingly, discovering the Higgs boson was not the first thing on their minds. More immediately, the detector offered the ATLAS team the clearest view yet of one of the twelve fundamental particles that the Standard Model has identified as the basic building blocks of matter. This particle, the top quark, is by far the heaviest of the twelve. With a mass of around 173 GeV, it even outweighs the newly discovered Higgs boson.

In the years ahead, the ATLAS detector will be tasked with examining both the top quark and the new Higgs boson in much more detail. Doing so could reveal whether the properties of these particles deviate in subtle ways from Standard Model predictions. If they do, physicists will have unambiguous evidence that their model isn't correct, and that new physics discoveries remain to be made. Those new discoveries could lead to new insights into the nature of gravity and the mysterious dark matter that we know makes up about 85% of the mass of the Universe, but that has so far evaded detection.

The ATLAS detector could do more than simply hint that such discoveries are possible, though – it could make some of them.

Hubble Space Telescope image of galaxy cluster Cl 0024+17 (a collision of two gigantic galaxy clusters). Dark matter in this cluster is detected by its strong gravitational effect on the light of galaxies far behind the cluster, an effect called gravitational lensing. The resulting dark matter distribution (shown as a "ghostly" bluish ring) is superimposed on the image of the cluster.

Turn up the energy

The physics experiments at ATLAS are well underway now. In 2010 and 2011, the two beams of protons colliding at the centre of the ATLAS detector each had a record-breaking 3.5 teraelectronvolts (or 3500 GeV) of energy. The beams collided at 7 TeV. By 2012 the LHC had broken its own record – the beams of protons were each 4 TeV, and the collisions 8 TeV. But there's more to come.

After a planned shutdown in 2013 and 2014, the LHC will resume operations in 2015 near to its full potential, with beams circulating close to their design energy of 7 TeV each. The window on new physics will be thrown wide open. What might ATLAS find in this unexplored country?

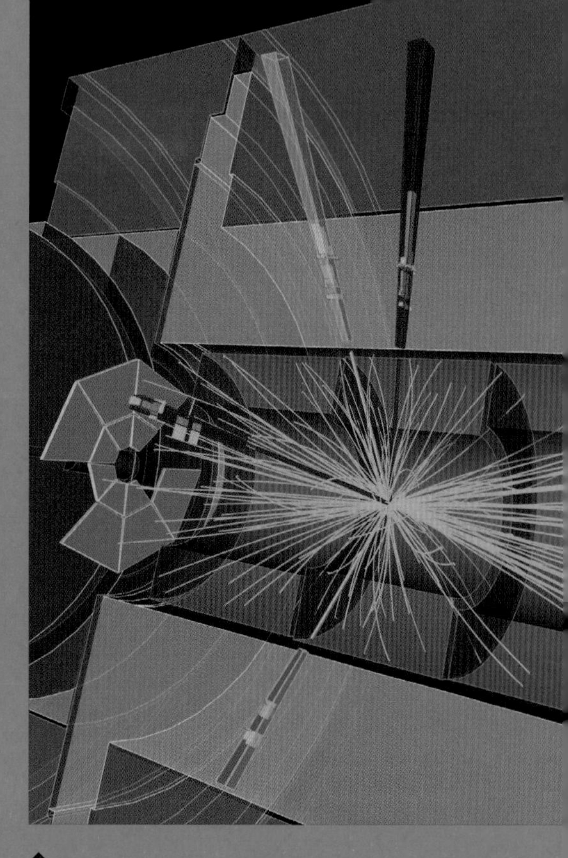

▲
Candidate event for a Higgs boson decaying into two pairs of electron, coloured red and blue.

The unbearable lightness of particles

The ATLAS detector has found a Higgs boson. With a mass of 125 GeV, it is among the heaviest particles ever found. The trouble is, it could have been much, much heavier.

The subatomic world crackles with quantum activity. This activity should exert an influence on the Higgs boson (as well as some of the other particle types within the subatomic world) that naturally renders the particle somewhere in the order of 10 quadrillion – 10,000,000,000,000,000 – times heavier than the mass of the Higgs boson discovered by the ATLAS team. Having a Higgs boson with a mass of just 125 GeV raises an important question: why is it so light? What are we missing in the theoretical framework that makes the Higgs boson mass immune to all this quantum activity?

Among the most elegant solutions to the light Higgs boson problem is to enlist a whole new set of particles, which behave almost like twins of the known elementary particles. This idea is known as Supersymmetry. Physicists have already named these hidden twins: the Higgs boson's twin has been dubbed the "Higgsino", for instance, while the heaviest

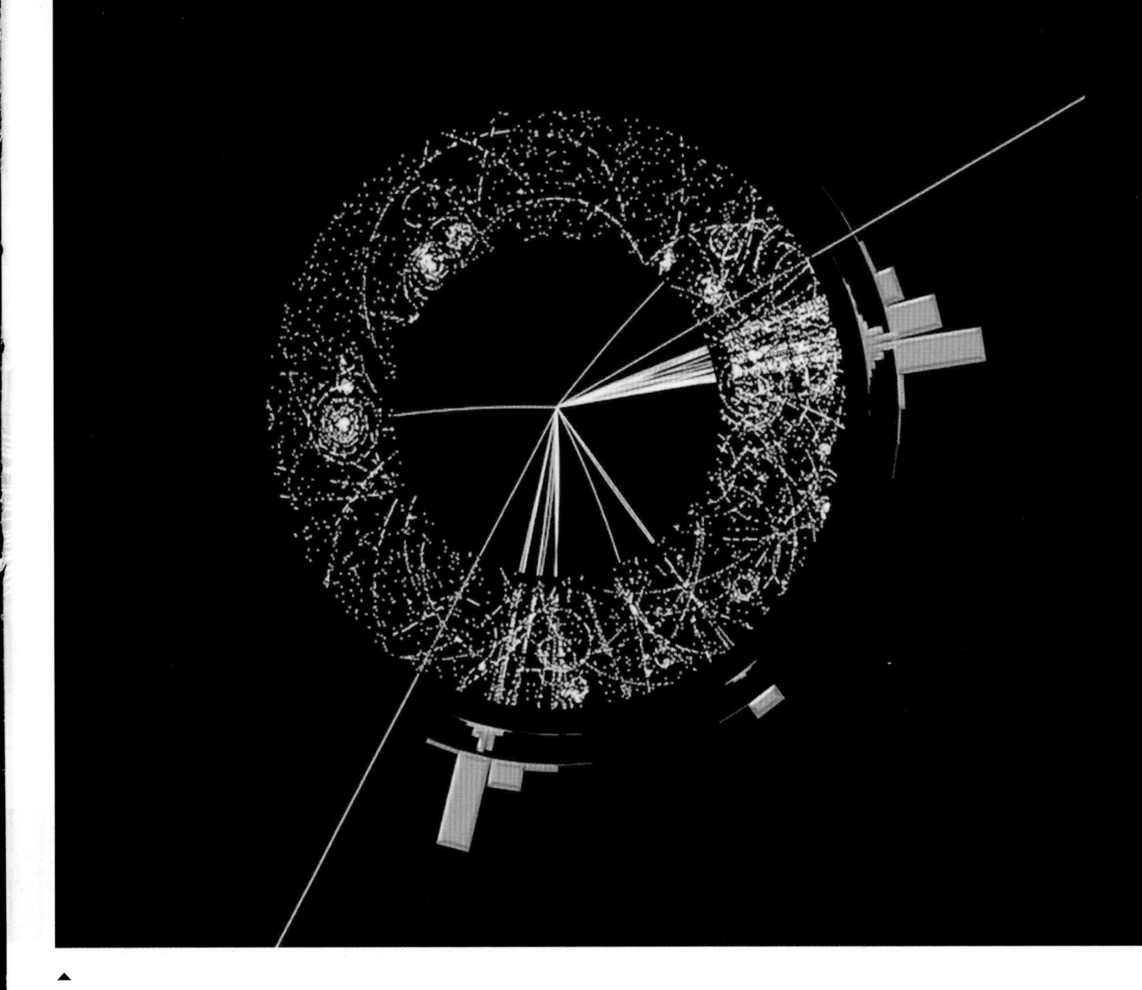

▲
Simulated supersymmetric event in the ATLAS detector.

known particle – the top quark – is twinned with the "top squark".

The low mass of the Higgs could reflect the way subatomic processes act on these normal and supersymmetric particles. In theory, each supersymmetric twin helps to cancel the effects of its corresponding normal particle, just as a positive number and its negative twin cancel when added. When all this cancelling out is taken into account, it leaves particles like the Higgs with masses far lower than the quantum processes acting on the particles would prefer.

The Higgsino and top squark – if they exist – may be too heavy for the ATLAS detector to discover. But Supersymmetry requires the twins of some Standard Model particles to be relatively light. Should they exist, they may be within reach of the ATLAS detector when it begins to search for particles created in collisions close to the design energy of 14 TeV. Finding these twins would be proof positive of new physics beyond the Standard Model, giving physicists the opportunity to explore some of the exciting properties such ideas entail.

Adding a new dimension

$$\delta_\varepsilon F = i\bar{\varepsilon}\bar{\sigma}^\mu \partial_\mu \psi$$

$$\delta_\varepsilon \psi_\alpha = i(\sigma^\mu \bar{\varepsilon})_\alpha \partial_\mu A + \varepsilon_\alpha F$$

$$\delta_\varepsilon A = \varepsilon \psi$$

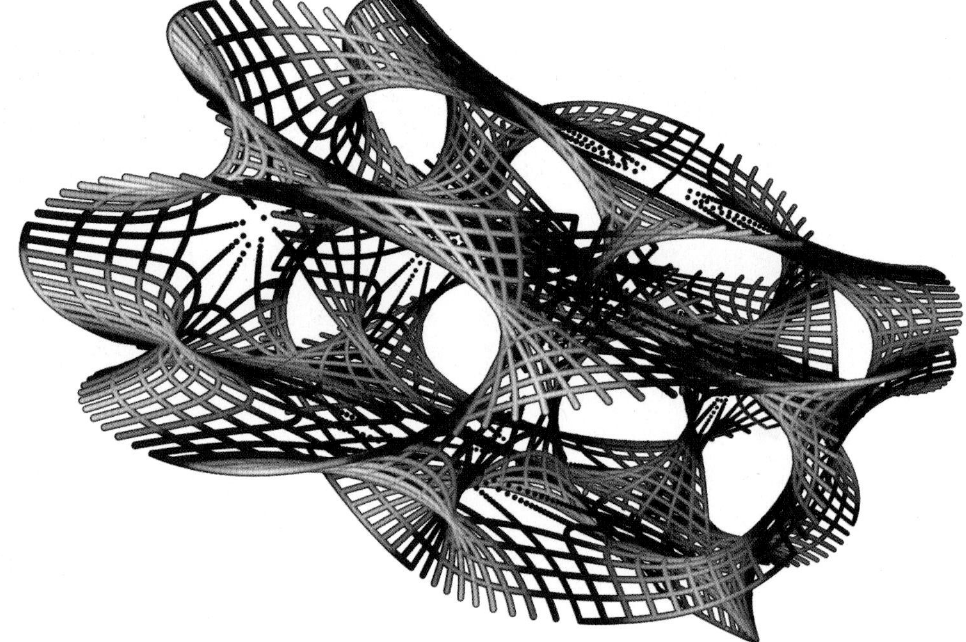

▲ A 3D rendered image of a 4-dimensional structure.

▶ Maintenance and upgrade work will take ATLAS further into the unknown world beyond the physics textbooks.

Supersymmetry is an idea older than the LHC project itself. But in the 40 years since it was first suggested, physicists have not yet found any direct evidence that known particles really do have missing twins. Luckily, there are already alternative ideas in place in case the ATLAS detector and the other LHC experiments continue to draw a blank in the search for Supersymmetry.

Another popular idea is that the Higgs owes its unexpected lightness not to extra particles but to extra dimensions. This idea relates to the apparent weakness of that most familiar of the four fundamental forces, the force of gravity.

To particle physicists, gravity is an enigma: this force is so weak compared to the strong, weak and electromagnetic forces that it hardly registers at all on the subatomic scale. That may be because gravity is actually strong, but acts largely in dimensions we have yet to discover, so we only see a weak 'echo' of its strength in the dimensions we are familiar with. It is just possible that as the ATLAS detector continues to probe the subatomic world, it might produce particles that travel in these extra dimensions as well as the four dimensions that we are familiar with. Evidence of these exotic particles and the strange dimensions in which they normally exist would give physicists another way to account for the Higgs boson's lightness.

Stepping into the unknown

Supersymmetry and extra dimensions are two of the most popular ideas physicists have devised to explain why some of their observations of the subatomic world seem to clash with theory. But the truth is that for particle physicists, there is much more to know than is already known, and it remains a mystery exactly what the ATLAS detector will find in the future.

Physicists have had many notable successes in their exploration of matter, but some things just don't add up. Most significantly, the way the Universe behaves only makes sense if it contains much more matter than physicists have managed to identify to date. In fact, their calculations suggest that a surprising 85% of the mass in the Universe is missing.

Like the hunt for the Higgs boson, the search for this mysterious missing matter – dark matter – has encouraged physicists to build and run all manner of experiments. So far, the search has been unsuccessful. Will the ATLAS detector enter the history books as the experiment that finally threw some much-needed light on dark matter?

With so many ideas and theories to test, the ATLAS team is going to be busy. The 7000-tonne ATLAS detector may be buried 100 metres below ground, but hopes are sky high that it is on the verge of more truly momentous discoveries. Whatever those discoveries may be, they promise to profoundly change the way we think about the Universe.

ATLAS
in numbers

27km

Circumference of LHC tunnel

46m long, 25m tall

Dimensions of ATLAS

100 Metres

Depth of the tunnel

-271°C

The LHC is one of the coldest objects in the Universe

7000 Tonnes

Weight of the detector

3000 Physicists

From 177 institutions in 38 countries are involved in the ATLAS Experiment

1 Billion

Approximate number of proton collisions per second

4 July 2012

Announcement of Higgs boson discovery

10 Billion km

Distance one LHC beam may travel in total as it circulates – the equivalent of a return journey to Neptune

2808

Maximum number of proton bunches within each beam

160 Billion

Number of protons within each bunch

11,245 Laps

Laps per second made by a proton in the LHC beam

7 TeV (Teraelectronvolts)

Maximum energy of each proton beam flying around the LHC

14 TeV

Maximum energy produced when two protons collide

3000 km

Length of wires and fibres within the ATLAS detector to carry information

125 Petabytes

Total data volume from ATLAS between 2010 and 2012

Chers hôtes,
Nous vous souhaitons **une bonne nuit**.

Voici les prévisions météorologiques pour demain.

Dear guest,
Have ***a good night sleep****.*

The weather forecast for tomorrow.

TEMPERATURE 14 °C 57 °F

☐ ☐ ☑ ☐ ☐ ☐